MULTIPLICATION
PRACTICE with 10-20

$$\begin{array}{r} 18 \\ \times 7 \\ \hline 126 \end{array} \qquad \begin{array}{r} 12 \\ \times 5 \\ \hline 60 \end{array} \qquad \begin{array}{r} 15 \\ \times 14 \\ \hline 210 \end{array}$$

Chris McMullen, Ph.D.

Multiplication Practice with 10-20
Improve Your Math Fluency Worksheets
Chris McMullen, Ph.D.

Copyright © 2019 Chris McMullen, Ph.D.

www.monkeyphysicsblog.wordpress.com
www.improveyourmathfluency.com
www.chrismcmullen.wordpress.com

Zishka Publishing
ISBN: 978-1-941691-51-9

Textbooks > Math > Arithmetic
Study Guides > Workbooks> Math
Education > Math > Arithmetic

CONTENTS

INTRODUCTION

This workbook is for students who have already learned to multiply 1 thru 9, and who are ready to advance to 10-20. (Students who need a quick review of the facts 1 thru 9 can find a review section on page 100.)

This workbook begins by focusing on one number (from 10 thru 20) at a time for a couple of pages. The exercises that follow multiply a one-digit number by a two-digit number. The mixed practice section includes everything from 1×1 thru 20×20. After this the workbook focuses on two-digit multiplication from 10×10 thru 20×20.

May you (or your students) find this workbook useful and become more fluent with multiplication.

Multiplication Table

	1	2	3	4	5	6	7	8	9	10	11	12	13	14	15	16	17	18	19	20
1	1	2	3	4	5	6	7	8	9	10	11	12	13	14	15	16	17	18	19	20
2	2	4	6	8	10	12	14	16	18	20	22	24	26	28	30	32	34	36	38	40
3	3	6	9	12	15	18	21	24	27	30	33	36	39	42	45	48	51	54	57	60
4	4	8	12	16	20	24	28	32	36	40	44	48	52	56	60	64	68	72	76	80
5	5	10	15	20	25	30	35	40	45	50	55	60	65	70	75	80	85	90	95	100
6	6	12	18	24	30	36	42	48	54	60	66	72	78	84	90	96	102	108	114	120
7	7	14	21	28	35	42	49	56	63	70	77	84	91	98	105	112	119	126	133	140
8	8	16	24	32	40	48	56	64	72	80	88	96	104	112	120	128	136	144	152	160
9	9	18	27	36	45	54	63	72	81	90	99	108	117	126	135	144	153	162	171	180
10	10	20	30	40	50	60	70	80	90	100	110	120	130	140	150	160	170	180	190	200
11	11	22	33	44	55	66	77	88	99	110	121	132	143	154	165	176	187	198	209	220
12	12	24	36	48	60	72	84	96	108	120	132	144	156	168	180	192	204	216	228	240
13	13	26	39	52	65	78	91	104	117	130	143	156	169	182	195	208	221	234	247	260
14	14	28	42	56	70	84	98	112	126	140	154	168	182	196	210	224	238	252	266	280
15	15	30	45	60	75	90	105	120	135	150	165	180	195	210	225	240	255	270	285	300
16	16	32	48	64	80	96	112	128	144	160	176	192	208	224	240	256	272	288	304	320
17	17	34	51	68	85	102	119	136	153	170	187	204	221	238	255	272	289	306	323	340
18	18	36	54	72	90	108	126	144	162	180	198	216	234	252	270	288	306	324	342	360
19	19	38	57	76	95	114	133	152	171	190	209	228	247	266	285	304	323	342	361	380
20	20	40	60	80	100	120	140	160	180	200	220	240	260	280	300	320	340	360	380	400

Example 1 (Carry-Over Method)

$$16$$
$$\times\,5$$

First multiply $6 \times 5 = 30$.
Write the 3 above the 1 and the 0 below the 5.

$$\overset{3}{1}6$$
$$\times\,5$$
$$\overline{0}$$

Now multiply $1 \times 5 = 5$.
Add the 3 from before to get $5 + 3 = 8$.
Write the 8 to the left of the zero.

$$\overset{3}{1}6$$
$$\times\,5$$
$$\overline{80}$$

The answer is $16 \times 5 = 80$.

Example 1 (Distributive Method)

$$16$$
$$\times\, 5$$
$$\overline{}$$

Note that $16 = 10 + 6$.

$$16 \times 5 = (10 + 6) \times 5 = 10 \times 5 + 6 \times 5$$

Write the answers to $6 \times 5 = 30$ and $10 \times 5 = 50$ on separate lines. Add $30 + 50 = 80$.

$$16$$
$$\times\, 5$$
$$\overline{}$$
$$30$$
$$\underline{50}$$
$$80$$

The answer is $16 \times 5 = 80$.

Example 1 (Table Method)

$$16$$
$$\times 5$$

	1	2	3	4	5	6	7	8	9	10	11	12	13	14	15	16	17	18	19	20
1	1	2	3	4	5	6	7	8	9	10	11	12	13	14	15	16	17	18	19	20
2	2	4	6	8	10	12	14	16	18	20	22	24	26	28	30	32	34	36	38	40
3	3	6	9	12	15	18	21	24	27	30	33	36	39	42	45	48	51	54	57	60
4	4	8	12	16	20	24	28	32	36	40	44	48	52	56	60	64	68	72	76	80
5	5	10	15	20	25	30	35	40	45	50	55	60	65	70	75	80	85	90	95	100
6	6	12	18	24	30	36	42	48	54	60	66	72	78	84	90	96	102	108	114	120
7	7	14	21	28	35	42	49	56	63	70	77	84	91	98	105	112	119	126	133	140
8	8	16	24	32	40	48	56	64	72	80	88	96	104	112	120	128	136	144	152	160
9	9	18	27	36	45	54	63	72	81	90	99	108	117	126	135	144	153	162	171	180
10	10	20	30	40	50	60	70	80	90	100	110	120	130	140	150	160	170	180	190	200
11	11	22	33	44	55	66	77	88	99	110	121	132	143	154	165	176	187	198	209	220
12	12	24	36	48	60	72	84	96	108	120	132	144	156	168	180	192	204	216	228	240
13	13	26	39	52	65	78	91	104	117	130	143	156	169	182	195	208	221	234	247	260
14	14	28	42	56	70	84	98	112	126	140	154	168	182	196	210	224	238	252	266	280
15	15	30	45	60	75	90	105	120	135	150	165	180	195	210	225	240	255	270	285	300
16	16	32	48	64	80	96	112	128	144	160	176	192	208	224	240	256	272	288	304	320
17	17	34	51	68	85	102	119	136	153	170	187	204	221	238	255	272	289	306	323	340
18	18	36	54	72	90	108	126	144	162	180	198	216	234	252	270	288	306	324	342	360
19	19	38	57	76	95	114	133	152	171	190	209	228	247	266	285	304	323	342	361	380
20	20	40	60	80	100	120	140	160	180	200	220	240	260	280	300	320	340	360	380	400

The answer is $16 \times 5 = 80$.

Example 2 (Carry-Over Method)

$$18$$
$$\times 16$$

First multiply $18 \times 6 = 108$ by following Example 1.

$$
\begin{array}{r}
\overset{4}{18} \\
\times 16 \\
\hline
108
\end{array}
$$

Next multiply $18 \times 10 = 180$. (Why is it 18×10 instead of 18×1? Because the 1 of **16** is in the tens place.)
Now add $108 + 180 = 288$.

$$
\begin{array}{r}
\overset{4}{18} \\
\times 16 \\
\hline
108 \\
180 \\
\hline
288
\end{array}
$$

The answer is $18 \times 16 = 288$.

Example 2 (Distributive Method)

$$18$$
$$\times 16$$

Note that $18 = 10 + 8$.

$$18 \times 16 = (10 + 8) \times 16 = 10 \times 16 + 8 \times 16$$

Following Example 1, note that $8 \times 16 = 8 \times 10 + 8 \times 6$. Write the answers to $10 \times 16 = 160$, $8 \times 10 = 80$, and $8 \times 6 = 48$ on separate lines. Add $160 + 80 + 48 = 288$.

$$18$$
$$\times 16$$
$$\overline{}$$
$$160$$
$$80$$
$$48$$
$$\overline{}$$
$$288$$

The answer is $18 \times 16 = 288$.

Example 2 (Table Method)

$$18$$
$$\times 16$$

	1	2	3	4	5	6	7	8	9	10	11	12	13	14	15	16	17	18	19	20
1	1	2	3	4	5	6	7	8	9	10	11	12	13	14	15	16	17	18	19	20
2	2	4	6	8	10	12	14	16	18	20	22	24	26	28	30	32	34	36	38	40
3	3	6	9	12	15	18	21	24	27	30	33	36	39	42	45	48	51	54	57	60
4	4	8	12	16	20	24	28	32	36	40	44	48	52	56	60	64	68	72	76	80
5	5	10	15	20	25	30	35	40	45	50	55	60	65	70	75	80	85	90	95	100
6	6	12	18	24	30	36	42	48	54	60	66	72	78	84	90	96	102	108	114	120
7	7	14	21	28	35	42	49	56	63	70	77	84	91	98	105	112	119	126	133	140
8	8	16	24	32	40	48	56	64	72	80	88	96	104	112	120	128	136	144	152	160
9	9	18	27	36	45	54	63	72	81	90	99	108	117	126	135	144	153	162	171	180
10	10	20	30	40	50	60	70	80	90	100	110	120	130	140	150	160	170	180	190	200
11	11	22	33	44	55	66	77	88	99	110	121	132	143	154	165	176	187	198	209	220
12	12	24	36	48	60	72	84	96	108	120	132	144	156	168	180	192	204	216	228	240
13	13	26	39	52	65	78	91	104	117	130	143	156	169	182	195	208	221	234	247	260
14	14	28	42	56	70	84	98	112	126	140	154	168	182	196	210	224	238	252	266	280
15	15	30	45	60	75	90	105	120	135	150	165	180	195	210	225	240	255	270	285	300
16	16	32	48	64	80	96	112	128	144	160	176	192	208	224	240	256	272	288	304	320
17	17	34	51	68	85	102	119	136	153	170	187	204	221	238	255	272	289	306	323	340
18	18	36	54	72	90	108	126	144	162	180	198	216	234	252	270	288	306	324	342	360
19	19	38	57	76	95	114	133	152	171	190	209	228	247	266	285	304	323	342	361	380
20	20	40	60	80	100	120	140	160	180	200	220	240	260	280	300	320	340	360	380	400

The answer is $18 \times 16 = 288$.

Practice with 10

① 10 × 8

② 17 ×10

③ 10 × 5

④ 20 ×10

⑤ 10 × 6

⑥ 16 ×10

⑦ 10 × 1

⑧ 11 ×10

⑨ 10 ×10

⑩ 15 ×10

⑪ 13 ×10

⑫ 10 × 3

⑬ 12 ×10

⑭ 10 × 4

⑮ 19 ×10

⑯ 10 × 9

⑰ 14 ×10

⑱ 10 × 2

⑲ 18 ×10

⑳ 10 × 7

㉑ 11 ×10

㉒ 10 × 2

㉓ 14 ×10

㉔ 10 ×10

㉕ 17 ×10

㉖ 10 × 7

㉗ 12 ×10

㉘ 10 × 8

㉙ 20 ×10

㉚ 10 × 3

㉛ 10 × 1

㉜ 15 ×10

㉝ 10 × 4

㉞ 16 ×10

㉟ 10 × 6

㊱ 19 ×10

㊲ 10 × 9

㊳ 13 ×10

㊴ 10 × 5

㊵ 18 ×10

Practice with 10

①
```
  17
×10
----
```
②
```
  10
× 6
----
```
③
```
  19
×10
----
```
④
```
  10
× 8
----
```
⑤
```
  20
×10
----
```
⑥
```
  10
× 7
----
```
⑦
```
  13
×10
----
```
⑧
```
  10
× 2
----
```
⑨
```
  16
×10
----
```
⑩
```
  10
×10
----
```

⑪
```
  14
×10
----
```
⑫
```
  10
× 4
----
```
⑬
```
  15
×10
----
```
⑭
```
  10
× 1
----
```
⑮
```
  16
×10
----
```
⑯
```
  10
× 5
----
```
⑰
```
  12
×10
----
```
⑱
```
  10
× 6
----
```
⑲
```
  17
×10
----
```
⑳
```
  10
× 3
----
```

㉑
```
  10
× 7
----
```
㉒
```
  11
×10
----
```
㉓
```
  10
×10
----
```
㉔
```
  13
×10
----
```
㉕
```
  10
× 9
----
```
㉖
```
  18
×10
----
```
㉗
```
  10
× 2
----
```
㉘
```
  20
×10
----
```
㉙
```
  10
× 8
----
```
㉚
```
  19
×10
----
```

㉛
```
  10
× 6
----
```
㉜
```
  19
×10
----
```
㉝
```
  10
× 3
----
```
㉞
```
  20
×10
----
```
㉟
```
  10
× 2
----
```
㊱
```
  15
×10
----
```
㊲
```
  10
× 1
----
```
㊳
```
  16
×10
----
```
㊴
```
  10
×10
----
```
㊵
```
  14
×10
----
```

Practice with 11

① 11 ② 15 ③ 11 ④ 12 ⑤ 11 ⑥ 14 ⑦ 11 ⑧ 13 ⑨ 11 ⑩ 11
× 5   ×11   × 9   ×11   ×10   ×11   × 7   ×11   × 2   ×11

⑪ 17 ⑫ 11 ⑬ 19 ⑭ 11 ⑮ 20 ⑯ 11 ⑰ 16 ⑱ 11 ⑲ 18 ⑳ 11
×11   × 1   ×11   × 3   ×11   × 6   ×11   × 4   ×11   × 8

㉑ 13 ㉒ 11 ㉓ 16 ㉔ 11 ㉕ 15 ㉖ 11 ㉗ 19 ㉘ 11 ㉙ 12 ㉚ 11
×11   × 4   ×11   × 2   ×11   × 8   ×11   × 5   ×11   × 1

㉛ 11 ㉜ 11 ㉝ 11 ㉞ 14 ㉟ 11 ㊱ 20 ㊲ 11 ㊳ 17 ㊴ 11 ㊵ 18
× 7   ×11   × 3   ×11   ×10   ×11   × 6   ×11   × 9   ×11

Practice with 11

① 15 ×11   ② 11 ×10   ③ 20 ×11   ④ 11 × 5   ⑤ 12 ×11   ⑥ 11 × 8   ⑦ 17 ×11   ⑧ 11 × 4   ⑨ 14 ×11   ⑩ 11 × 2

⑪ 16 ×11   ⑫ 11 × 3   ⑬ 11 ×11   ⑭ 11 × 7   ⑮ 14 ×11   ⑯ 11 × 9   ⑰ 19 ×11   ⑱ 11 ×10   ⑲ 15 ×11   ⑳ 11 × 1

㉑ 11 × 8   ㉒ 13 ×11   ㉓ 11 × 2   ㉔ 17 ×11   ㉕ 11 × 6   ㉖ 18 ×11   ㉗ 11 × 4   ㉘ 12 ×11   ㉙ 11 × 5   ㉚ 20 ×11

㉛ 11 ×10   ㉜ 20 ×11   ㉝ 11 × 1   ㉞ 12 ×11   ㉟ 11 × 4   ㊱ 11 ×11   ㊲ 11 × 7   ㊳ 14 ×11   ㊴ 11 × 2   ㊵ 16 ×11

Practice with 12

① 12 × 6

② 19 ×12

③ 12 × 8

④ 13 ×12

⑤ 12 × 3

⑥ 20 ×12

⑦ 12 × 5

⑧ 14 ×12

⑨ 12 × 4

⑩ 15 ×12

⑪ 12 ×12

⑫ 12 × 2

⑬ 11 ×12

⑭ 12 × 7

⑮ 18 ×12

⑯ 12 × 1

⑰ 16 ×12

⑱ 12 × 9

⑲ 17 ×12

⑳ 12 ×10

㉑ 14 ×12

㉒ 12 × 9

㉓ 16 ×12

㉔ 12 × 4

㉕ 19 ×12

㉖ 12 ×10

㉗ 11 ×12

㉘ 12 × 6

㉙ 13 ×12

㉚ 12 × 2

㉛ 12 × 5

㉜ 15 ×12

㉝ 12 × 7

㉞ 20 ×12

㉟ 12 × 3

㊱ 18 ×12

㊲ 12 × 1

㊳ 12 ×12

㊴ 12 × 8

㊵ 17 ×12

Practice with 12

① 19 ×12
② 12 × 3
③ 18 ×12
④ 12 × 6
⑤ 13 ×12
⑥ 12 ×10
⑦ 12 ×12
⑧ 12 × 9
⑨ 20 ×12
⑩ 12 × 4

⑪ 16 ×12
⑫ 12 × 7
⑬ 15 ×12
⑭ 12 × 5
⑮ 20 ×12
⑯ 12 × 8
⑰ 11 ×12
⑱ 12 × 3
⑲ 19 ×12
⑳ 12 × 2

㉑ 12 ×10
㉒ 14 ×12
㉓ 12 × 4
㉔ 12 ×12
㉕ 12 × 1
㉖ 17 ×12
㉗ 12 × 9
㉘ 13 ×12
㉙ 12 × 6
㉚ 18 ×12

㉛ 12 × 3
㉜ 18 ×12
㉝ 12 × 2
㉞ 13 ×12
㉟ 12 × 9
㊱ 15 ×12
㊲ 12 × 5
㊳ 20 ×12
㊴ 12 × 4
㊵ 16 ×12

Practice with 13

① 13
× 4

② 18
×13

③ 13
×10

④ 14
×13

⑤ 13
× 6

⑥ 20
×13

⑦ 13
× 7

⑧ 15
×13

⑨ 13
× 8

⑩ 16
×13

⑪ 17
×13

⑫ 13
× 1

⑬ 12
×13

⑭ 13
× 2

⑮ 19
×13

⑯ 13
× 5

⑰ 11
×13

⑱ 13
× 3

⑲ 13
×13

⑳ 13
× 9

㉑ 15
×13

㉒ 13
× 3

㉓ 11
×13

㉔ 13
× 8

㉕ 18
×13

㉖ 13
× 9

㉗ 12
×13

㉘ 13
× 4

㉙ 14
×13

㉚ 13
× 1

㉛ 13
× 7

㉜ 16
×13

㉝ 13
× 2

㉞ 20
×13

㉟ 13
× 6

㊱ 19
×13

㊲ 13
× 5

㊳ 17
×13

㊴ 13
×10

㊵ 13
×13

Practice with 13

① 18　② 13　③ 19　④ 13　⑤ 14　⑥ 13　⑦ 17　⑧ 13　⑨ 20　⑩ 13
　×13　　× 6　　×13　　× 4　　×13　　× 9　　×13　　× 3　　×13　　× 8

⑪ 11　⑫ 13　⑬ 16　⑭ 13　⑮ 20　⑯ 13　⑰ 12　⑱ 13　⑲ 18　⑳ 13
　×13　　× 2　　×13　　× 7　　×13　　×10　　×13　　× 6　　×13　　× 1

㉑ 13　㉒ 15　㉓ 13　㉔ 17　㉕ 13　㉖ 13　㉗ 13　㉘ 14　㉙ 13　㉚ 19
　× 9　　×13　　× 8　　×13　　× 5　　×13　　× 3　　×13　　× 4　　×13

㉛ 13　㉜ 19　㉝ 13　㉞ 14　㉟ 13　㊱ 16　㊲ 13　㊳ 20　㊴ 13　㊵ 11
　× 6　　×13　　× 1　　×13　　× 3　　×13　　× 7　　×13　　× 8　　×13

Practice with 14

①
14
× 1
___

②
12
×14
___

③
14
× 3
___

④
18
×14
___

⑤
14
× 2
___

⑥
19
×14
___

⑦
14
× 5
___

⑧
16
×14
___

⑨
14
× 9
___

⑩
13
×14
___

⑪
14
×14
___

⑫
14
× 8
___

⑬
17
×14
___

⑭
14
×10
___

⑮
15
×14
___

⑯
14
× 6
___

⑰
11
×14
___

⑱
14
× 4
___

⑲
20
×14
___

⑳
14
× 7
___

㉑
16
×14
___

㉒
14
× 4
___

㉓
11
×14
___

㉔
14
× 9
___

㉕
12
×14
___

㉖
14
× 7
___

㉗
17
×14
___

㉘
14
× 1
___

㉙
18
×14
___

㉚
14
× 8
___

㉛
14
× 5
___

㉜
13
×14
___

㉝
14
×10
___

㉞
19
×14
___

㉟
14
× 2
___

㊱
15
×14
___

㊲
14
× 6
___

㊳
14
×14
___

㊴
14
× 3
___

㊵
20
×14
___

Practice with 14

① 12 ×14
② 14 × 2
③ 15 ×14
④ 14 × 1
⑤ 18 ×14
⑥ 14 × 7
⑦ 14 ×14
⑧ 14 × 4
⑨ 19 ×14
⑩ 14 × 9

⑪ 11 ×14
⑫ 14 ×10
⑬ 13 ×14
⑭ 14 × 5
⑮ 19 ×14
⑯ 14 × 3
⑰ 17 ×14
⑱ 14 × 2
⑲ 12 ×14
⑳ 14 × 8

㉑ 14 × 7
㉒ 16 ×14
㉓ 14 × 9
㉔ 14 ×14
㉕ 14 × 6
㉖ 20 ×14
㉗ 14 × 4
㉘ 18 ×14
㉙ 14 × 1
㉚ 15 ×14

㉛ 14 × 2
㉜ 15 ×14
㉝ 14 × 8
㉞ 18 ×14
㉟ 14 × 4
㊱ 13 ×14
㊲ 14 × 5
㊳ 19 ×14
㊴ 14 × 9
㊵ 11 ×14

Practice with 15

①
15
× 9
---

②
15
×15
---

③
15
× 3
---

④
14
×15
---

⑤
15
× 2
---

⑥
19
×15
---

⑦
15
× 1
---

⑧
17
×15
---

⑨
15
× 6
---

⑩
13
×15
---

⑪
12
×15
---

⑫
15
×10
---

⑬
16
×15
---

⑭
15
× 8
---

⑮
18
×15
---

⑯
15
× 7
---

⑰
11
×15
---

⑱
15
× 5
---

⑲
20
×15
---

⑳
15
× 4

㉑
17
×15

㉒
15
× 5

㉓
11
×15

㉔
15
× 6

㉕
15
×15

㉖
15
× 4

㉗
16
×15

㉘
15
× 9

㉙
14
×15

㉚
15
×10

㉛
15
× 1

㉜
13
×15

㉝
15
× 8

㉞
19
×15

㉟
15
× 2

㊱
18
×15

㊲
15
× 7

㊳
12
×15

㊴
15
× 3

㊵
20
×15

Practice with 15

①   ②   ③   ④   ⑤   ⑥   ⑦   ⑧   ⑨   ⑩

```
   15      15      18      15      14      15      12      15      19      15
  ×15    ×  2    ×15    ×  9    ×15    ×  4    ×15    ×  5    ×15    ×  6
```

⑪   ⑫   ⑬   ⑭   ⑮   ⑯   ⑰   ⑱   ⑲   ⑳

```
   11      15      13      15      19      15      16      15      15      15
  ×15    ×  8    ×15    ×  1    ×15    ×  3    ×15    ×  2    ×15    ×10
```

㉑   ㉒   ㉓   ㉔   ㉕   ㉖   ㉗   ㉘   ㉙   ㉚

```
   15      17      15      12      15      20      15      14      15      18
  ×  4    ×15    ×  6    ×15    ×  7    ×15    ×  5    ×15    ×  9    ×15
```

㉛   ㉜   ㉝   ㉞   ㉟   ㊱   ㊲   ㊳   ㊴   ㊵

```
   15      18      15      14      15      13      15      19      15      11
  ×  2    ×15    ×10    ×15    ×  5    ×15    ×  1    ×15    ×  6    ×15
```

Practice with 16

① 16 × 2 ② 18 ×16 ③ 16 × 6 ④ 11 ×16 ⑤ 16 × 1 ⑥ 16 ×16 ⑦ 16 × 7 ⑧ 17 ×16 ⑨ 16 × 4 ⑩ 15 ×16

⑪ 20 ×16 ⑫ 16 ×10 ⑬ 19 ×16 ⑭ 16 × 8 ⑮ 12 ×16 ⑯ 16 × 3 ⑰ 13 ×16 ⑱ 16 × 5 ⑲ 14 ×16 ⑳ 16 × 9

㉑ 17 ×16 ㉒ 16 × 5 ㉓ 13 ×16 ㉔ 16 × 4 ㉕ 18 ×16 ㉖ 16 × 9 ㉗ 19 ×16 ㉘ 16 × 2 ㉙ 11 ×16 ㉚ 16 ×10

㉛ 16 × 7 ㉜ 15 ×16 ㉝ 16 × 8 ㉞ 16 ×16 ㉟ 16 × 1 ㊱ 12 ×16 ㊲ 16 × 3 ㊳ 20 ×16 ㊴ 16 × 6 ㊵ 14 ×16

Practice with 16

① 18 ×16

② 16 × 1

③ 12 ×16

④ 16 × 2

⑤ 11 ×16

⑥ 16 × 9

⑦ 20 ×16

⑧ 16 × 5

⑨ 16 ×16

⑩ 16 × 4

⑪ 13 ×16

⑫ 16 × 8

⑬ 15 ×16

⑭ 16 × 7

⑮ 16 ×16

⑯ 16 × 6

⑰ 19 ×16

⑱ 16 × 1

⑲ 18 ×16

⑳ 16 ×10

㉑ 16 × 9

㉒ 17 ×16

㉓ 16 × 4

㉔ 20 ×16

㉕ 16 × 3

㉖ 14 ×16

㉗ 16 × 5

㉘ 11 ×16

㉙ 16 × 2

㉚ 12 ×16

㉛ 16 × 1

㉜ 12 ×16

㉝ 16 ×10

㉞ 11 ×16

㉟ 16 × 5

㊱ 15 ×16

㊲ 16 × 7

㊳ 16 ×16

㊴ 16 × 4

㊵ 13 ×16

Practice with 17

①   ②   ③   ④   ⑤   ⑥   ⑦   ⑧   ⑨   ⑩
  17   11   17   19   17   12   17   16   17   18
×10   ×17   × 5   ×17   × 6   ×17   × 1   ×17   × 2   ×17

⑪   ⑫   ⑬   ⑭   ⑮   ⑯   ⑰   ⑱   ⑲   ⑳
  15   17   20   17   14   17   13   17   17   17
×17   × 8   ×17   × 4   ×17   × 9   ×17   × 7   ×17   × 3

㉑   ㉒   ㉓   ㉔   ㉕   ㉖   ㉗   ㉘   ㉙   ㉚
  16   17   13   17   11   17   20   17   19   17
×17   × 7   ×17   × 2   ×17   × 3   ×17   ×10   ×17   × 8

㉛   ㉜   ㉝   ㉞   ㉟   ㊱   ㊲   ㊳   ㊴   ㊵
  17   18   17   12   17   14   17   15   17   17
× 1   ×17   × 4   ×17   × 6   ×17   × 9   ×17   × 5   ×17

Practice with 17

① 11 ×17

② 17 × 6

③ 14 ×17

④ 17 ×10

⑤ 19 ×17

⑥ 17 × 3

⑦ 15 ×17

⑧ 17 × 7

⑨ 12 ×17

⑩ 17 × 2

⑪ 13 ×17

⑫ 17 × 4

⑬ 18 ×17

⑭ 17 × 1

⑮ 12 ×17

⑯ 17 × 5

⑰ 20 ×17

⑱ 17 × 6

⑲ 11 ×17

⑳ 17 × 8

㉑ 17 × 3

㉒ 16 ×17

㉓ 17 × 2

㉔ 15 ×17

㉕ 17 × 9

㉖ 17 ×17

㉗ 17 × 7

㉘ 19 ×17

㉙ 17 ×10

㉚ 14 ×17

㉛ 17 × 6

㉜ 14 ×17

㉝ 17 × 8

㉞ 19 ×17

㉟ 17 × 7

㊱ 18 ×17

㊲ 17 × 1

㊳ 12 ×17

㊴ 17 × 2

㊵ 13 ×17

Practice with 18

① 18 × 7

② 11 ×18

③ 18 × 1

④ 12 ×18

⑤ 18 ×10

⑥ 20 ×18

⑦ 18 × 6

⑧ 15 ×18

⑨ 18 × 8

⑩ 14 ×18

⑪ 18 ×18

⑫ 18 × 3

⑬ 16 ×18

⑭ 18 × 2

⑮ 13 ×18

⑯ 18 × 4

⑰ 17 ×18

⑱ 18 × 5

⑲ 19 ×18

⑳ 18 × 9

㉑ 15 ×18

㉒ 18 × 5

㉓ 17 ×18

㉔ 18 × 8

㉕ 11 ×18

㉖ 18 × 9

㉗ 16 ×18

㉘ 18 × 7

㉙ 12 ×18

㉚ 18 × 3

㉛ 18 × 6

㉜ 14 ×18

㉝ 18 × 2

㉞ 20 ×18

㉟ 18 ×10

㊱ 13 ×18

㊲ 18 × 4

㊳ 18 ×18

㊴ 18 × 1

㊵ 19 ×18

Practice with 18

① 11 ×18

② 18 ×10

③ 13 ×18

④ 18 × 7

⑤ 12 ×18

⑥ 18 × 9

⑦ 18 ×18

⑧ 18 × 5

⑨ 20 ×18

⑩ 18 × 8

⑪ 17 ×18

⑫ 18 × 2

⑬ 14 ×18

⑭ 18 × 6

⑮ 20 ×18

⑯ 18 × 1

⑰ 16 ×18

⑱ 18 ×10

⑲ 11 ×18

⑳ 18 × 3

㉑ 18 × 9

㉒ 15 ×18

㉓ 18 × 8

㉔ 18 ×18

㉕ 18 × 4

㉖ 19 ×18

㉗ 18 × 5

㉘ 12 ×18

㉙ 18 × 7

㉚ 13 ×18

㉛ 18 ×10

㉜ 13 ×18

㉝ 18 × 3

㉞ 12 ×18

㉟ 18 × 5

㊱ 14 ×18

㊲ 18 × 6

㊳ 20 ×18

㊴ 18 × 8

㊵ 17 ×18

Practice with 19

① 　　② 　　③ 　　④ 　　⑤ 　　⑥ 　　⑦ 　　⑧ 　　⑨ 　　⑩

```
   19     20     19     17     19     13     19     16     19     15
×  3   ×19    ×  9   ×19    ×  2   ×19    ×  4   ×19    ×  7   ×19
```

⑪ 　　⑫ 　　⑬ 　　⑭ 　　⑮ 　　⑯ 　　⑰ 　　⑱ 　　⑲ 　　⑳

```
   12     19     18     19     11     19     14     19     19     19
×19    ×10    ×19    ×  8   ×19    ×  6   ×19    ×  1   ×19    ×  5
```

㉑ 　　㉒ 　　㉓ 　　㉔ 　　㉕ 　　㉖ 　　㉗ 　　㉘ 　　㉙ 　　㉚

```
   16     19     14     19     20     19     18     19     17     19
×19    ×  1   ×19    ×  7   ×19    ×  5   ×19    ×  3   ×19    ×10
```

㉛ 　　㉜ 　　㉝ 　　㉞ 　　㉟ 　　㊱ 　　㊲ 　　㊳ 　　㊴ 　　㊵

```
   19     15     19     13     19     11     19     12     19     19
×  4   ×19    ×  8   ×19    ×  2   ×19    ×  6   ×19    ×  9   ×19
```

Practice with 19

① 20 ×19
② 19 × 2
③ 11 ×19
④ 19 × 3
⑤ 17 ×19
⑥ 19 × 5
⑦ 12 ×19
⑧ 19 × 1
⑨ 13 ×19
⑩ 19 × 7

⑪ 14 ×19
⑫ 19 × 8
⑬ 15 ×19
⑭ 19 × 4
⑮ 13 ×19
⑯ 19 × 9
⑰ 18 ×19
⑱ 19 × 2
⑲ 20 ×19
⑳ 19 ×10

㉑ 19 × 5
㉒ 16 ×19
㉓ 19 × 7
㉔ 12 ×19
㉕ 19 × 6
㉖ 19 ×19
㉗ 19 × 1
㉘ 17 ×19
㉙ 19 × 3
㉚ 11 ×19

㉛ 19 × 2
㉜ 11 ×19
㉝ 19 ×10
㉞ 17 ×19
㉟ 19 × 1
㊱ 15 ×19
㊲ 19 × 4
㊳ 13 ×19
㊴ 19 × 7
㊵ 14 ×19

Practice with 20

① 20 × 6
② 19 ×20
③ 20 × 2
④ 13 ×20
⑤ 20 × 8
⑥ 15 ×20
⑦ 20 × 5
⑧ 18 ×20
⑨ 20 × 1
⑩ 20 ×20

⑪ 16 ×20
⑫ 20 ×10
⑬ 14 ×20
⑭ 20 × 4
⑮ 17 ×20
⑯ 20 × 3
⑰ 12 ×20
⑱ 20 × 9
⑲ 11 ×20
⑳ 20 × 7

㉑ 18 ×20
㉒ 20 × 9
㉓ 12 ×20
㉔ 20 × 1
㉕ 19 ×20
㉖ 20 × 7
㉗ 14 ×20
㉘ 20 × 6
㉙ 13 ×20
㉚ 20 ×10

㉛ 20 × 5
㉜ 20 ×20
㉝ 20 × 4
㉞ 15 ×20
㉟ 20 × 8
㊱ 17 ×20
㊲ 20 × 3
㊳ 16 ×20
㊴ 20 × 2
㊵ 11 ×20

Practice with 20

① 19 ×20 ② 20 × 8 ③ 17 ×20 ④ 20 × 6 ⑤ 13 ×20 ⑥ 20 × 7 ⑦ 16 ×20 ⑧ 20 × 9 ⑨ 15 ×20 ⑩ 20 × 1

⑪ 12 ×20 ⑫ 20 × 4 ⑬ 20 ×20 ⑭ 20 × 5 ⑮ 15 ×20 ⑯ 20 × 2 ⑰ 14 ×20 ⑱ 20 × 8 ⑲ 19 ×20 ⑳ 20 ×10

㉑ 20 × 7 ㉒ 18 ×20 ㉓ 20 × 1 ㉔ 16 ×20 ㉕ 20 × 3 ㉖ 11 ×20 ㉗ 20 × 9 ㉘ 13 ×20 ㉙ 20 × 6 ㉚ 17 ×20

㉛ 20 × 8 ㉜ 17 ×20 ㉝ 20 ×10 ㉞ 13 ×20 ㉟ 20 × 9 ㊱ 20 ×20 ㊲ 20 × 5 ㊳ 15 ×20 ㊴ 20 × 1 ㊵ 12 ×20

Two Digits Times One Digit

①	②	③	④	⑤	⑥	⑦	⑧	⑨	⑩
16	17	20	15	18	12	14	19	13	11
× 8	× 1	× 6	× 5	× 7	× 4	× 2	× 8	× 9	× 3

⑪	⑫	⑬	⑭	⑮	⑯	⑰	⑱	⑲	⑳
12	14	11	19	13	10	20	16	18	17
× 9	× 6	× 2	× 5	× 1	× 5	× 4	× 3	× 8	× 7

㉑	㉒	㉓	㉔	㉕	㉖	㉗	㉘	㉙	㉚
20	15	10	16	19	11	18	17	14	13
× 9	× 2	× 3	× 1	× 3	× 6	× 5	× 8	× 4	× 7

㉛	㉜	㉝	㉞	㉟	㊱	㊲	㊳	㊴	㊵
11	18	17	20	14	13	15	16	10	12
× 4	× 3	× 5	× 1	× 9	× 8	× 6	× 2	× 2	× 7

Two Digits Times One Digit

①	②	③	④	⑤	⑥	⑦	⑧	⑨	⑩
14	10	13	18	12	16	19	20	11	15
× 1	× 6	× 5	× 2	× 8	× 3	× 6	× 7	× 9	× 4

⑪	⑫	⑬	⑭	⑮	⑯	⑰	⑱	⑲	⑳
16	20	12	11	15	19	13	10	17	14
× 6	× 8	× 5	× 1	× 9	× 4	× 3	× 4	× 2	× 7

㉑	㉒	㉓	㉔	㉕	㉖	㉗	㉘	㉙	㉚
13	19	15	10	17	14	12	16	20	18
× 2	× 9	× 1	× 9	× 6	× 8	× 3	× 7	× 5	× 4

㉛	㉜	㉝	㉞	㉟	㊱	㊲	㊳	㊴	㊵
17	12	14	13	11	20	19	18	15	10
× 4	× 2	× 5	× 6	× 8	× 3	× 1	× 9	× 7	× 1

Two Digits Times One Digit

① 11 × 4   ② 10 × 7   ③ 18 × 2   ④ 15 × 8   ⑤ 17 × 9   ⑥ 14 × 5   ⑦ 19 × 3   ⑧ 13 × 4   ⑨ 20 × 6   ⑩ 12 × 1

⑪ 14 × 6   ⑫ 19 × 2   ⑬ 12 × 3   ⑭ 13 × 8   ⑮ 20 × 7   ⑯ 16 × 8   ⑰ 18 × 5   ⑱ 11 × 1   ⑲ 17 × 4   ⑳ 10 × 9

㉑ 18 × 6   ㉒ 15 × 3   ㉓ 16 × 1   ㉔ 11 × 7   ㉕ 13 × 1   ㉖ 12 × 2   ㉗ 17 × 8   ㉘ 10 × 4   ㉙ 19 × 5   ㉚ 20 × 9

㉛ 12 × 5   ㉜ 17 × 1   ㉝ 10 × 8   ㉞ 18 × 7   ㉟ 19 × 6   ㊱ 20 × 4   ㊲ 15 × 2   ㊳ 11 × 3   ㊴ 16 × 3   ㊵ 14 × 9

Two Digits Times One Digit

① 19   ② 16   ③ 20   ④ 17   ⑤ 14   ⑥ 11   ⑦ 13   ⑧ 18   ⑨ 12   ⑩ 15
× 7   × 2   × 8   × 3   × 4   × 1   × 2   × 9   × 6   × 5

⑪ 11   ⑫ 18   ⑬ 14   ⑭ 12   ⑮ 15   ⑯ 13   ⑰ 20   ⑱ 16   ⑲ 10   ⑳ 19
× 2   × 4   × 8   × 7   × 6   × 5   × 1   × 5   × 3   × 9

㉑ 20   ㉒ 13   ㉓ 15   ㉔ 16   ㉕ 10   ㉖ 19   ㉗ 14   ㉘ 11   ㉙ 18   ㉚ 17
× 3   × 6   × 7   × 6   × 2   × 4   × 1   × 9   × 8   × 5

㉛ 10   ㉜ 14   ㉝ 19   ㉞ 20   ㉟ 12   ㊱ 18   ㊲ 13   ㊳ 17   ㊴ 15   ㊵ 16
× 5   × 3   × 8   × 2   × 4   × 1   × 7   × 6   × 9   × 7

Two Digits Times One Digit

① 18   ② 15   ③ 11   ④ 19   ⑤ 13   ⑥ 14   ⑦ 12   ⑧ 10   ⑨ 20   ⑩ 17
× 4   × 6   × 2   × 1   × 7   × 5   × 8   × 4   × 3   × 9
────   ────   ────   ────   ────   ────   ────   ────   ────   ────

⑪ 14   ⑫ 12   ⑬ 17   ⑭ 10   ⑮ 20   ⑯ 16   ⑰ 11   ⑱ 18   ⑲ 13   ⑳ 15
× 3   × 2   × 8   × 1   × 6   × 1   × 5   × 9   × 4   × 7
────   ────   ────   ────   ────   ────   ────   ────   ────   ────

㉑ 11   ㉒ 19   ㉓ 16   ㉔ 18   ㉕ 10   ㉖ 17   ㉗ 13   ㉘ 15   ㉙ 12   ㉚ 20
× 3   × 8   × 9   × 6   × 9   × 2   × 1   × 4   × 5   × 7
────   ────   ────   ────   ────   ────   ────   ────   ────   ────

㉛ 17   ㉜ 13   ㉝ 15   ㉞ 11   ㉟ 12   ㊱ 20   ㊲ 19   ㊳ 18   ㊴ 16   ㊵ 14
× 5   × 9   × 1   × 6   × 3   × 4   × 2   × 8   × 8   × 7
────   ────   ────   ────   ────   ────   ────   ────   ────   ────

Two Digits Times One Digit

① 12 × 6
② 16 × 2
③ 20 × 1
④ 13 × 8
⑤ 14 × 4
⑥ 18 × 9
⑦ 10 × 2
⑧ 11 × 7
⑨ 17 × 3
⑩ 19 × 5

⑪ 18 × 2
⑫ 11 × 4
⑬ 14 × 1
⑭ 17 × 6
⑮ 19 × 3
⑯ 10 × 5
⑰ 20 × 9
⑱ 16 × 5
⑲ 15 × 8
⑳ 12 × 7

㉑ 20 × 8
㉒ 10 × 3
㉓ 19 × 6
㉔ 16 × 3
㉕ 15 × 2
㉖ 12 × 4
㉗ 14 × 9
㉘ 18 × 7
㉙ 11 × 1
㉚ 13 × 5

㉛ 15 × 5
㉜ 14 × 8
㉝ 12 × 1
㉞ 20 × 2
㉟ 17 × 4
㊱ 11 × 9
㊲ 10 × 6
㊳ 13 × 3
㊴ 19 × 7
㊵ 16 × 6

Two Digits Times One Digit

①	②	③	④	⑤	⑥	⑦	⑧	⑨	⑩
18	19	12	17	16	14	10	15	20	13
× 8	× 2	× 3	× 5	× 6	× 1	× 4	× 8	× 9	× 7

⑪	⑫	⑬	⑭	⑮	⑯	⑰	⑱	⑲	⑳
14	10	13	15	20	11	12	18	16	19
× 9	× 3	× 4	× 5	× 2	× 5	× 1	× 7	× 8	× 6

㉑	㉒	㉓	㉔	㉕	㉖	㉗	㉘	㉙	㉚
12	17	11	18	15	13	16	19	10	20
× 9	× 4	× 7	× 2	× 7	× 3	× 5	× 8	× 1	× 6

㉛	㉜	㉝	㉞	㉟	㊱	㊲	㊳	㊴	㊵
13	16	19	12	10	20	17	18	11	14
× 1	× 7	× 5	× 2	× 9	× 8	× 3	× 4	× 4	× 6

____________________ ________ ________ _____ / 40

(name) (date) (time) (score)

Two Digits Times One Digit

①	②	③	④	⑤	⑥	⑦	⑧	⑨	⑩
10	11	20	16	14	18	15	12	13	17
× 2	× 3	× 5	× 4	× 8	× 7	× 3	× 6	× 9	× 1
⑪	⑫	⑬	⑭	⑮	⑯	⑰	⑱	⑲	⑳
18	12	14	13	17	15	20	11	19	10
× 3	× 8	× 5	× 2	× 9	× 1	× 7	× 1	× 4	× 6
㉑	㉒	㉓	㉔	㉕	㉖	㉗	㉘	㉙	㉚
20	15	17	11	19	10	14	18	12	16
× 4	× 9	× 2	× 9	× 3	× 8	× 7	× 6	× 5	× 1
㉛	㉜	㉝	㉞	㉟	㊱	㊲	㊳	㊴	㊵
19	14	10	20	13	12	15	16	17	11
× 1	× 4	× 5	× 3	× 8	× 7	× 2	× 9	× 6	× 2

Two Digits Times One Digit

①	②	③	④	⑤	⑥	⑦	⑧	⑨	⑩
20	13	17	19	16	18	12	10	11	14
× 3	× 2	× 5	× 1	× 4	× 8	× 7	× 3	× 9	× 6

⑪	⑫	⑬	⑭	⑮	⑯	⑰	⑱	⑲	⑳
18	12	14	10	11	15	17	20	16	13
× 9	× 5	× 7	× 1	× 2	× 1	× 8	× 6	× 3	× 4

㉑	㉒	㉓	㉔	㉕	㉖	㉗	㉘	㉙	㉚
17	19	15	20	10	14	16	13	12	11
× 9	× 7	× 6	× 2	× 6	× 5	× 1	× 3	× 8	× 4

㉛	㉜	㉝	㉞	㉟	㊱	㊲	㊳	㊴	㊵
14	16	13	17	12	11	19	20	15	18
× 8	× 6	× 1	× 2	× 9	× 3	× 5	× 7	× 7	× 4

Two Digits Times One Digit

① 12 × 2

② 15 × 5

③ 11 × 1

④ 16 × 7

⑤ 18 × 3

⑥ 20 × 6

⑦ 10 × 5

⑧ 17 × 4

⑨ 14 × 9

⑩ 19 × 8

⑪ 20 × 5

⑫ 17 × 3

⑬ 18 × 1

⑭ 14 × 2

⑮ 19 × 9

⑯ 10 × 8

⑰ 11 × 6

⑱ 15 × 8

⑲ 13 × 7

⑳ 12 × 4

㉑ 11 × 7

㉒ 10 × 9

㉓ 19 × 2

㉔ 15 × 9

㉕ 13 × 5

㉖ 12 × 3

㉗ 18 × 6

㉘ 20 × 4

㉙ 17 × 1

㉚ 16 × 8

㉛ 13 × 8

㉜ 18 × 7

㉝ 12 × 1

㉞ 11 × 5

㉟ 14 × 3

㊱ 17 × 6

㊲ 10 × 2

㊳ 16 × 9

㊴ 19 × 4

㊵ 15 × 2

Two Digits Times One Digit

① 12 × 2
② 17 × 5
③ 10 × 1
④ 18 × 6
⑤ 16 × 3
⑥ 19 × 4
⑦ 20 × 7
⑧ 11 × 2
⑨ 13 × 8
⑩ 14 × 9

⑪ 19 × 8
⑫ 20 × 1
⑬ 14 × 7
⑭ 11 × 6
⑮ 13 × 5
⑯ 15 × 6
⑰ 10 × 4
⑱ 12 × 9
⑲ 16 × 2
⑳ 17 × 3

㉑ 10 × 8
㉒ 18 × 7
㉓ 15 × 9
㉔ 12 × 5
㉕ 11 × 9
㉖ 14 × 1
㉗ 16 × 6
㉘ 17 × 2
㉙ 20 × 4
㉚ 13 × 3

㉛ 14 × 4
㉜ 16 × 9
㉝ 17 × 6
㉞ 10 × 5
㉟ 20 × 8
㊱ 13 × 2
㊲ 18 × 1
㊳ 12 × 7
㊴ 15 × 7
㊵ 19 × 3

Two Digits Times One Digit

① 20 × 5
② 15 × 1
③ 13 × 6
④ 16 × 7
⑤ 19 × 2
⑥ 12 × 9
⑦ 11 × 1
⑧ 10 × 3
⑨ 14 × 8
⑩ 18 × 4

⑪ 12 × 1
⑫ 10 × 2
⑬ 19 × 6
⑭ 14 × 5
⑮ 18 × 8
⑯ 11 × 4
⑰ 13 × 9
⑱ 15 × 4
⑲ 17 × 7
⑳ 20 × 3

㉑ 13 × 7
㉒ 11 × 8
㉓ 18 × 5
㉔ 15 × 8
㉕ 17 × 1
㉖ 20 × 2
㉗ 19 × 9
㉘ 12 × 3
㉙ 10 × 6
㉚ 16 × 4

㉛ 17 × 4
㉜ 19 × 7
㉝ 20 × 6
㉞ 13 × 1
㉟ 14 × 2
㊱ 10 × 9
㊲ 11 × 5
㊳ 16 × 8
㊴ 18 × 3
㊵ 15 × 5

Two Digits Times One Digit

① 17 × 6   ② 10 × 4   ③ 14 × 2   ④ 20 × 7   ⑤ 12 × 9   ⑥ 15 × 1   ⑦ 11 × 8   ⑧ 16 × 6   ⑨ 18 × 3   ⑩ 13 × 5

⑪ 15 × 3   ⑫ 11 × 2   ⑬ 13 × 8   ⑭ 16 × 7   ⑮ 18 × 4   ⑯ 19 × 7   ⑰ 14 × 1   ⑱ 17 × 5   ⑲ 12 × 6   ⑳ 10 × 9

㉑ 14 × 3   ㉒ 20 × 8   ㉓ 19 × 5   ㉔ 17 × 4   ㉕ 16 × 5   ㉖ 13 × 2   ㉗ 12 × 7   ㉘ 10 × 6   ㉙ 11 × 1   ㉚ 18 × 9

㉛ 13 × 1   ㉜ 12 × 5   ㉝ 10 × 7   ㉞ 14 × 4   ㉟ 11 × 3   ㊱ 18 × 6   ㊲ 20 × 2   ㊳ 17 × 8   ㊴ 19 × 8   ㊵ 15 × 9

Two Digits Times One Digit

①	②	③	④	⑤	⑥	⑦	⑧	⑨	⑩
11	19	18	12	15	17	16	14	13	20
× 4	× 2	× 7	× 8	× 6	× 5	× 2	× 9	× 3	× 1

⑪	⑫	⑬	⑭	⑮	⑯	⑰	⑱	⑲	⑳
17	14	15	13	20	16	18	19	10	11
× 2	× 6	× 7	× 4	× 3	× 1	× 5	× 1	× 8	× 9

㉑	㉒	㉓	㉔	㉕	㉖	㉗	㉘	㉙	㉚
18	16	20	19	10	11	15	17	14	12
× 8	× 3	× 4	× 3	× 2	× 6	× 5	× 9	× 7	× 1

㉛	㉜	㉝	㉞	㉟	㊱	㊲	㊳	㊴	㊵
10	15	11	18	13	14	16	12	20	19
× 1	× 8	× 7	× 2	× 6	× 5	× 4	× 3	× 9	× 4

Two Digits Times One Digit

①	②	③	④	⑤	⑥	⑦	⑧	⑨	⑩
12	10	13	14	19	11	20	15	17	18
× 5	× 9	× 4	× 8	× 1	× 2	× 7	× 5	× 6	× 3

⑪	⑫	⑬	⑭	⑮	⑯	⑰	⑱	⑲	⑳
11	20	18	15	17	16	13	12	19	10
× 6	× 4	× 7	× 8	× 9	× 8	× 2	× 3	× 5	× 1

㉑	㉒	㉓	㉔	㉕	㉖	㉗	㉘	㉙	㉚
13	14	16	12	15	18	19	10	20	17
× 6	× 7	× 3	× 9	× 3	× 4	× 8	× 5	× 2	× 1

㉛	㉜	㉝	㉞	㉟	㊱	㊲	㊳	㊴	㊵
18	19	10	13	20	17	14	12	16	11
× 2	× 3	× 8	× 9	× 6	× 5	× 4	× 7	× 7	× 1

Two Digits Times One Digit

① 20 × 9   ② 16 × 4   ③ 17 × 8   ④ 19 × 7   ⑤ 11 × 5   ⑥ 12 × 3   ⑦ 15 × 4   ⑧ 13 × 1   ⑨ 18 × 6   ⑩ 14 × 2

⑪ 12 × 4   ⑫ 13 × 5   ⑬ 11 × 8   ⑭ 18 × 9   ⑮ 14 × 6   ⑯ 15 × 2   ⑰ 17 × 3   ⑱ 16 × 2   ⑲ 10 × 7   ⑳ 20 × 1

㉑ 17 × 7   ㉒ 15 × 6   ㉓ 14 × 9   ㉔ 16 × 6   ㉕ 10 × 4   ㉖ 20 × 5   ㉗ 11 × 3   ㉘ 12 × 1   ㉙ 13 × 8   ㉚ 19 × 2

㉛ 10 × 2   ㉜ 11 × 7   ㉝ 20 × 8   ㉞ 17 × 4   ㉟ 18 × 5   ㊱ 13 × 3   ㊲ 15 × 9   ㊳ 19 × 6   ㊴ 14 × 1   ㊵ 16 × 9

Two Digits Times One Digit

①	②	③	④	⑤	⑥	⑦	⑧	⑨	⑩
19	10	11	12	15	18	14	16	17	20
× 9	× 1	× 2	× 7	× 6	× 3	× 8	× 9	× 4	× 5

⑪	⑫	⑬	⑭	⑮	⑯	⑰	⑱	⑲	⑳
18	14	20	16	17	13	11	19	15	10
× 4	× 2	× 8	× 7	× 1	× 7	× 3	× 5	× 9	× 6

㉑	㉒	㉓	㉔	㉕	㉖	㉗	㉘	㉙	㉚
11	12	13	19	16	20	15	10	14	17
× 4	× 8	× 5	× 1	× 5	× 2	× 7	× 9	× 3	× 6

㉛	㉜	㉝	㉞	㉟	㊱	㊲	㊳	㊴	㊵
20	15	10	11	14	17	12	19	13	18
× 3	× 5	× 7	× 1	× 4	× 9	× 2	× 8	× 8	× 6

Two Digits Times One Digit

① 14   ② 13   ③ 17   ④ 15   ⑤ 18   ⑥ 19   ⑦ 16   ⑧ 11   ⑨ 20   ⑩ 12
× 1   × 2   × 7   × 8   × 9   × 5   × 2   × 6   × 4   × 3

⑪ 19   ⑫ 11   ⑬ 18   ⑭ 20   ⑮ 12   ⑯ 16   ⑰ 17   ⑱ 13   ⑲ 10   ⑳ 14
× 2   × 9   × 7   × 1   × 4   × 3   × 5   × 3   × 8   × 6

㉑ 17   ㉒ 16   ㉓ 12   ㉔ 13   ㉕ 10   ㉖ 14   ㉗ 18   ㉘ 19   ㉙ 11   ㉚ 15
× 8   × 4   × 1   × 4   × 2   × 9   × 5   × 6   × 7   × 3

㉛ 10   ㉜ 18   ㉝ 14   ㉞ 17   ㉟ 20   ㊱ 11   ㊲ 16   ㊳ 15   ㊴ 12   ㊵ 13
× 3   × 8   × 7   × 2   × 9   × 5   × 1   × 4   × 6   × 1

Two Digits Times One Digit

① 19 × 8

② 17 × 5

③ 14 × 6

④ 11 × 9

⑤ 16 × 1

⑥ 18 × 7

⑦ 13 × 3

⑧ 15 × 8

⑨ 10 × 4

⑩ 20 × 2

⑪ 18 × 4

⑫ 13 × 6

⑬ 20 × 3

⑭ 15 × 9

⑮ 10 × 5

⑯ 12 × 9

⑰ 14 × 7

⑱ 19 × 2

⑲ 16 × 8

⑳ 17 × 1

㉑ 14 × 4

㉒ 11 × 3

㉓ 12 × 2

㉔ 19 × 5

㉕ 15 × 2

㉖ 20 × 6

㉗ 16 × 9

㉘ 17 × 8

㉙ 13 × 7

㉚ 10 × 1

㉛ 20 × 7

㉜ 16 × 2

㉝ 17 × 9

㉞ 14 × 5

㉟ 13 × 4

㊱ 10 × 8

㊲ 11 × 6

㊳ 19 × 3

㊴ 12 × 3

㊵ 18 × 1

Two Digits Times One Digit

① 13 × 5   ② 12 × 6   ③ 10 × 9   ④ 16 × 3   ⑤ 18 × 8   ⑥ 19 × 2   ⑦ 15 × 6   ⑧ 14 × 1   ⑨ 20 × 4   ⑩ 11 × 7

⑪ 19 × 6   ⑫ 14 × 8   ⑬ 18 × 9   ⑭ 20 × 5   ⑮ 11 × 4   ⑯ 15 × 7   ⑰ 10 × 2   ⑱ 12 × 7   ⑲ 17 × 3   ⑳ 13 × 1

㉑ 10 × 3   ㉒ 15 × 4   ㉓ 11 × 5   ㉔ 12 × 4   ㉕ 17 × 6   ㉖ 13 × 8   ㉗ 18 × 2   ㉘ 19 × 1   ㉙ 14 × 9   ㉚ 16 × 7

㉛ 17 × 7   ㉜ 18 × 3   ㉝ 13 × 9   ㉞ 10 × 6   ㉟ 20 × 8   ㊱ 14 × 2   ㊲ 15 × 5   ㊳ 16 × 4   ㊴ 11 × 1   ㊵ 12 × 5

Mixed Practice

①	②	③	④	⑤	⑥	⑦	⑧	⑨	⑩
10	16	14	19	12	11	13	15	20	18
× 4	×16	× 8	×12	× 3	×17	× 9	×19	× 5	×18

⑪	⑫	⑬	⑭	⑮	⑯	⑰	⑱	⑲	⑳
19	12	11	15	13	14	20	18	16	17
× 6	×13	×10	×11	× 1	×15	× 7	×14	× 2	×20

㉑	㉒	㉓	㉔	㉕	㉖	㉗	㉘	㉙	㉚
18	20	13	10	16	19	12	17	15	11
× 9	×12	× 5	×19	× 4	×18	× 8	×17	× 3	×16

㉛	㉜	㉝	㉞	㉟	㊱	㊲	㊳	㊴	㊵
15	11	16	13	20	18	17	14	10	12
× 7	×13	× 6	×20	×10	×11	× 2	×14	× 1	×15

______________________ (name) ________ (date) ________ (time) _____ / 40 (score)

Mixed Practice

① 13 × 8	② 10 ×18	③ 18 × 4	④ 12 ×16	⑤ 14 × 9	⑥ 16 ×12	⑦ 11 × 3	⑧ 19 ×17	⑨ 17 × 5	⑩ 20 ×19
⑪ 20 × 2	⑫ 14 ×11	⑬ 19 × 1	⑭ 17 ×13	⑮ 11 × 6	⑯ 15 ×14	⑰ 10 ×10	⑱ 16 ×20	⑲ 18 × 7	⑳ 13 ×15
㉑ 12 × 4	㉒ 18 ×17	㉓ 17 × 3	㉔ 14 ×19	㉕ 10 × 5	㉖ 20 ×16	㉗ 16 × 9	㉘ 13 ×12	㉙ 11 × 8	㉚ 15 ×18
㉛ 17 × 7	㉜ 19 ×13	㉝ 12 ×10	㉞ 16 ×15	㉟ 15 × 1	㊱ 13 ×14	㊲ 18 × 2	㊳ 11 ×11	㊴ 14 × 6	㊵ 10 ×20

Mixed Practice

① 14 × 7

② 11 ×14

③ 20 × 5

④ 17 ×17

⑤ 13 ×10

⑥ 15 ×18

⑦ 10 × 8

⑧ 12 ×11

⑨ 18 × 9

⑩ 19 ×15

⑪ 17 × 2

⑫ 13 ×12

⑬ 15 × 1

⑭ 12 ×16

⑮ 10 × 4

⑯ 20 ×13

⑰ 18 × 6

⑱ 19 ×19

⑲ 11 × 3

⑳ 16 ×20

㉑ 19 × 8

㉒ 18 ×17

㉓ 10 × 9

㉔ 14 ×11

㉕ 11 × 7

㉖ 17 ×15

㉗ 13 × 5

㉘ 16 ×18

㉙ 12 ×10

㉚ 15 ×14

㉛ 12 × 6

㉜ 15 ×12

㉝ 11 × 2

㉞ 10 ×20

㉟ 18 × 1

㊱ 19 ×16

㊲ 16 × 3

㊳ 20 ×19

㊴ 14 × 4

㊵ 13 ×13

Mixed Practice

① 10 × 5   ② 14 ×15   ③ 19 × 7   ④ 13 ×14   ⑤ 20 × 8   ⑥ 11 ×17   ⑦ 15 ×10   ⑧ 17 ×18   ⑨ 16 × 9   ⑩ 18 ×11

⑪ 18 × 3   ⑫ 20 ×16   ⑬ 17 × 4   ⑭ 16 ×12   ⑮ 15 × 2   ⑯ 12 ×19   ⑰ 14 × 1   ⑱ 11 ×20   ⑲ 19 × 6   ⑳ 10 ×13

㉑ 13 × 7   ㉒ 19 ×18   ㉓ 16 ×10   ㉔ 20 ×11   ㉕ 14 × 9   ㉖ 18 ×14   ㉗ 11 × 8   ㉘ 10 ×17   ㉙ 15 × 5   ㉚ 12 ×15

㉛ 16 × 6   ㉜ 17 ×12   ㉝ 13 × 1   ㉞ 11 ×13   ㉟ 12 × 4   ㊱ 10 ×19   ㊲ 19 × 3   ㊳ 15 ×16   ㊴ 20 × 2   ㊵ 14 ×20

Mixed Practice

①   ②   ③   ④   ⑤   ⑥   ⑦   ⑧   ⑨   ⑩
  19   18   20   16   11   10   12   13   14   15
×10   ×15   × 8   ×19   × 7   ×13   × 5   ×12   × 3   ×18

⑪   ⑫   ⑬   ⑭   ⑮   ⑯   ⑰   ⑱   ⑲   ⑳
  16   11   10   13   12   20   14   15   18   17
× 9   ×14   × 6   ×17   × 2   ×16   × 4   ×20   × 1   ×11

㉑   ㉒   ㉓   ㉔   ㉕   ㉖   ㉗   ㉘   ㉙   ㉚
  15   14   12   19   18   16   11   17   13   10
× 5   ×19   × 3   ×12   ×10   ×18   × 8   ×13   × 7   ×15

㉛   ㉜   ㉝   ㉞   ㉟   ㊱   ㊲   ㊳   ㊴   ㊵
  13   10   18   12   14   15   17   20   19   11
× 4   ×14   × 9   ×11   × 6   ×17   × 1   ×20   × 2   ×16

Mixed Practice

① 12 × 8
② 19 ×18
③ 15 ×10
④ 11 ×15
⑤ 20 × 5
⑥ 18 ×19
⑦ 10 × 7
⑧ 16 ×13
⑨ 17 × 3
⑩ 14 ×12

⑪ 14 × 1
⑫ 20 ×17
⑬ 16 × 2
⑭ 17 ×14
⑮ 10 × 9
⑯ 13 ×20
⑰ 19 × 6
⑱ 18 ×11
⑲ 15 × 4
⑳ 12 ×16

㉑ 11 ×10
㉒ 15 ×13
㉓ 17 × 7
㉔ 20 ×12
㉕ 19 × 3
㉖ 14 ×15
㉗ 18 × 5
㉘ 12 ×19
㉙ 10 × 8
㉚ 13 ×18

㉛ 17 × 4
㉜ 16 ×14
㉝ 11 × 6
㉞ 18 ×16
㉟ 13 × 2
㊱ 12 ×20
㊲ 15 × 1
㊳ 10 ×17
㊴ 20 × 9
㊵ 19 ×11

Mixed Practice

① 10   ② 20   ③ 14   ④ 18   ⑤ 13   ⑥ 19   ⑦ 15   ⑧ 17   ⑨ 11   ⑩ 16
  × 9   ×18   × 3   ×19   ×10   ×14   × 1   ×12   × 5   ×15

⑪ 18   ⑫ 13   ⑬ 19   ⑭ 17   ⑮ 15   ⑯ 14   ⑰ 11   ⑱ 16   ⑲ 20   ⑳ 12
  × 4   ×20   × 7   ×16   × 6   ×13   × 2   ×17   × 8   ×11

㉑ 16   ㉒ 11   ㉓ 15   ㉔ 10   ㉕ 20   ㉖ 18   ㉗ 13   ㉘ 12   ㉙ 17   ㉚ 19
  × 1   ×19   × 5   ×12   × 9   ×15   × 3   ×14   ×10   ×18

㉛ 17   ㉜ 19   ㉝ 20   ㉞ 15   ㉟ 11   ㊱ 16   ㊲ 12   ㊳ 14   ㊴ 10   ㊵ 13
  × 2   ×20   × 4   ×11   × 7   ×16   × 8   ×17   × 6   ×13

Mixed Practice

① 15 × 3   ② 10 ×15   ③ 16 × 9   ④ 13 ×18   ⑤ 14 × 1   ⑥ 20 ×19   ⑦ 19 ×10   ⑧ 18 ×14   ⑨ 12 × 5   ⑩ 11 ×12

⑪ 11 × 8   ⑫ 14 ×16   ⑬ 18 × 6   ⑭ 12 ×20   ⑮ 19 × 4   ⑯ 17 ×17   ⑰ 10 × 7   ⑱ 20 ×11   ⑲ 16 × 2   ⑳ 15 ×13

㉑ 13 × 9   ㉒ 16 ×14   ㉓ 12 ×10   ㉔ 14 ×12   ㉕ 10 × 5   ㉖ 11 ×18   ㉗ 20 × 1   ㉘ 15 ×19   ㉙ 19 × 3   ㉚ 17 ×15

㉛ 12 × 2   ㉜ 18 ×20   ㉝ 13 × 7   ㉞ 20 ×13   ㉟ 17 × 6   ㊱ 15 ×17   ㊲ 16 × 8   ㊳ 19 ×16   ㊴ 14 × 4   ㊵ 10 ×11

_____ _____ _____ ____ / 40
(name)   (date)   (time)   (score)

Mixed Practice

①	②	③	④	⑤	⑥	⑦	⑧	⑨	⑩
15	17	10	12	14	18	11	19	20	16
× 6	×18	× 7	×16	× 4	×15	× 9	×11	× 3	×17

⑪	⑫	⑬	⑭	⑮	⑯	⑰	⑱	⑲	⑳
12	14	18	19	11	10	20	16	17	13
× 1	×12	× 2	×14	×10	×19	× 5	×13	× 8	×20

㉑	㉒	㉓	㉔	㉕	㉖	㉗	㉘	㉙	㉚
16	20	11	15	17	12	14	13	19	18
× 9	×16	× 3	×11	× 6	×17	× 7	×15	× 4	×18

㉛	㉜	㉝	㉞	㉟	㊱	㊲	㊳	㊴	㊵
19	18	17	11	20	16	13	10	15	14
× 5	×12	× 1	×20	× 2	×14	× 8	×13	×10	×19

Mixed Practice

① 11   ② 15   ③ 16   ④ 14   ⑤ 10   ⑥ 17   ⑦ 18   ⑧ 12   ⑨ 13   ⑩ 20
× 7   ×17   × 6   ×18   × 9   ×16   × 4   ×15   × 3   ×11

⑪ 20   ⑫ 10   ⑬ 12   ⑭ 13   ⑮ 18   ⑯ 19   ⑰ 15   ⑱ 17   ⑲ 16   ⑳ 11
× 8   ×14   ×10   ×12   × 1   ×13   × 2   ×20   × 5   ×19

㉑ 14   ㉒ 16   ㉓ 13   ㉔ 10   ㉕ 15   ㉖ 20   ㉗ 17   ㉘ 11   ㉙ 18   ㉚ 19
× 6   ×15   × 4   ×11   × 3   ×18   × 9   ×16   × 7   ×17

㉛ 13   ㉜ 12   ㉝ 14   ㉞ 17   ㉟ 19   ㊱ 11   ㊲ 16   ㊳ 18   ㊴ 10   ㊵ 15
× 5   ×12   × 2   ×19   ×10   ×13   × 8   ×14   × 1   ×20

Mixed Practice

① 16 ② 14 ③ 10 ④ 11 ⑤ 19 ⑥ 17 ⑦ 18 ⑧ 12 ⑨ 20 ⑩ 13
× 8   ×20   × 5   ×16   × 2   ×14   ×10   ×17   × 3   ×19

⑪ 11 ⑫ 19 ⑬ 17 ⑭ 12 ⑮ 18 ⑯ 10 ⑰ 20 ⑱ 13 ⑲ 14 ⑳ 15
× 1   ×11   × 7   ×12   × 4   ×18   × 6   ×15   × 9   ×13

㉑ 13 ㉒ 20 ㉓ 18 ㉔ 16 ㉕ 14 ㉖ 11 ㉗ 19 ㉘ 15 ㉙ 12 ㉚ 17
×10   ×16   × 3   ×17   × 8   ×19   × 5   ×14   × 2   ×20

㉛ 12 ㉜ 17 ㉝ 14 ㉞ 18 ㉟ 20 ㊱ 13 ㊲ 15 ㊳ 10 ㊴ 16 ㊵ 19
× 6   ×11   × 1   ×13   × 7   ×12   × 9   ×15   × 4   ×18

Mixed Practice

① 18 × 5   ② 16 ×19   ③ 13 × 8   ④ 19 ×20   ⑤ 10 ×10   ⑥ 14 ×16   ⑦ 17 × 2   ⑧ 11 ×14   ⑨ 15 × 3   ⑩ 20 ×17

⑪ 20 × 9   ⑫ 10 ×12   ⑬ 11 × 4   ⑭ 15 ×11   ⑮ 17 × 1   ⑯ 12 ×15   ⑰ 16 × 7   ⑱ 14 ×13   ⑲ 13 × 6   ⑳ 18 ×18

㉑ 19 × 8   ㉒ 13 ×14   ㉓ 15 × 2   ㉔ 10 ×17   ㉕ 16 × 3   ㉖ 20 ×20   ㉗ 14 ×10   ㉘ 18 ×16   ㉙ 17 × 5   ㉚ 12 ×19

㉛ 15 × 6   ㉜ 11 ×11   ㉝ 19 × 7   ㉞ 14 ×18   ㉟ 12 × 4   ㊱ 18 ×15   ㊲ 13 × 9   ㊳ 17 ×12   ㊴ 10 × 1   ㊵ 16 ×13

Mixed Practice

① 17 × 8
② 12 ×18
③ 16 × 2
④ 11 ×14
⑤ 13 × 6
⑥ 20 ×13
⑦ 14 × 5
⑧ 18 ×12
⑨ 15 ×10
⑩ 10 ×20

⑪ 11 × 4
⑫ 13 ×17
⑬ 20 × 9
⑭ 18 ×19
⑮ 14 × 1
⑯ 16 ×15
⑰ 15 × 7
⑱ 10 ×16
⑲ 12 × 3
⑳ 19 ×11

㉑ 10 × 5
㉒ 15 ×14
㉓ 14 ×10
㉔ 17 ×12
㉕ 12 × 8
㉖ 11 ×20
㉗ 13 × 2
㉘ 19 ×13
㉙ 18 × 6
㉚ 20 ×18

㉛ 18 × 7
㉜ 20 ×17
㉝ 12 × 4
㉞ 14 ×11
㉟ 15 × 9
㊱ 10 ×19
㊲ 19 × 3
㊳ 16 ×16
㊴ 17 × 1
㊵ 13 ×15

Mixed Practice

① 14   ② 17   ③ 10   ④ 13   ⑤ 16   ⑥ 12   ⑦ 20   ⑧ 11   ⑨ 19   ⑩ 15
× 2   ×20   × 8   ×18   × 5   ×14   × 6   ×13   ×10   ×12

⑪ 15   ⑫ 16   ⑬ 11   ⑭ 19   ⑮ 20   ⑯ 18   ⑰ 17   ⑱ 12   ⑲ 10   ⑳ 14
× 3   ×19   × 1   ×17   × 4   ×16   × 9   ×11   × 7   ×15

㉑ 13   ㉒ 10   ㉓ 19   ㉔ 16   ㉕ 17   ㉖ 15   ㉗ 12   ㉘ 14   ㉙ 20   ㉚ 18
× 8   ×13   × 6   ×12   ×10   ×18   × 5   ×14   × 2   ×20

㉛ 19   ㉜ 11   ㉝ 13   ㉞ 12   ㉟ 18   ㊱ 14   ㊲ 10   ㊳ 20   ㊴ 16   ㊵ 17
× 7   ×17   × 9   ×15   × 1   ×16   × 3   ×19   × 4   ×11

Mixed Practice

① 12 × 1

② 10 ×15

③ 19 × 7

④ 14 ×14

⑤ 11 × 2

⑥ 20 ×16

⑦ 18 × 9

⑧ 13 ×13

⑨ 16 × 4

⑩ 17 ×18

⑪ 14 × 8

⑫ 11 ×11

⑬ 20 ×10

⑭ 13 ×12

⑮ 18 × 3

⑯ 19 ×20

⑰ 16 × 5

⑱ 17 ×19

⑲ 10 × 6

⑳ 15 ×17

㉑ 17 × 9

㉒ 16 ×14

㉓ 18 × 4

㉔ 12 ×13

㉕ 10 × 1

㉖ 14 ×18

㉗ 11 × 7

㉘ 15 ×16

㉙ 13 × 2

㉚ 20 ×15

㉛ 13 × 5

㉜ 20 ×11

㉝ 10 × 8

㉞ 18 ×17

㉟ 16 ×10

㊱ 17 ×12

㊲ 15 × 6

㊳ 19 ×19

㊴ 12 × 3

㊵ 11 ×20

Mixed Practice

①	②	③	④	⑤	⑥	⑦	⑧	⑨	⑩
18	12	17	11	19	10	20	14	15	16
× 7	×18	× 1	×15	× 9	×14	× 2	×16	× 4	×13

⑪	⑫	⑬	⑭	⑮	⑯	⑰	⑱	⑲	⑳
16	19	14	15	20	13	12	10	17	18
× 6	×12	× 3	×11	× 8	×19	×10	×17	× 5	×20

㉑	㉒	㉓	㉔	㉕	㉖	㉗	㉘	㉙	㉚
11	17	15	19	12	16	10	18	20	13
× 1	×16	× 2	×13	× 4	×15	× 9	×14	× 7	×18

㉛	㉜	㉝	㉞	㉟	㊱	㊲	㊳	㊴	㊵
15	14	11	10	13	18	17	20	19	12
× 5	×11	×10	×20	× 3	×19	× 6	×12	× 8	×17

Mixed Practice

① 20 × 8 ② 16 ×14 ③ 18 × 5 ④ 19 ×15 ⑤ 10 × 3 ⑥ 17 ×20 ⑦ 11 × 2 ⑧ 14 ×12 ⑨ 12 × 9 ⑩ 15 ×11

⑪ 19 × 1 ⑫ 10 ×18 ⑬ 17 ×10 ⑭ 14 ×19 ⑮ 11 × 7 ⑯ 18 ×16 ⑰ 12 × 6 ⑱ 15 ×17 ⑲ 16 × 4 ⑳ 13 ×13

㉑ 15 × 2 ㉒ 12 ×15 ㉓ 11 × 9 ㉔ 20 ×12 ㉕ 16 × 8 ㉖ 19 ×11 ㉗ 10 × 5 ㉘ 13 ×20 ㉙ 14 × 3 ㉚ 17 ×14

㉛ 14 × 6 ㉜ 17 ×18 ㉝ 16 × 1 ㉞ 11 ×13 ㉟ 12 ×10 ㊱ 15 ×19 ㊲ 13 × 4 ㊳ 18 ×17 ㊴ 20 × 7 ㊵ 10 ×16

Mixed Practice

| ① 11 × 5 | ② 20 ×11 | ③ 15 × 8 | ④ 10 ×14 | ⑤ 18 × 2 | ⑥ 16 ×15 | ⑦ 17 × 3 | ⑧ 19 ×20 | ⑨ 13 × 9 | ⑩ 12 ×12 |

| ⑪ 12 × 4 | ⑫ 18 ×19 | ⑬ 19 × 7 | ⑭ 13 ×18 | ⑮ 17 × 1 | ⑯ 14 ×17 | ⑰ 20 ×10 | ⑱ 16 ×13 | ⑲ 15 × 6 | ⑳ 11 ×16 |

| ㉑ 10 × 8 | ㉒ 15 ×20 | ㉓ 13 × 3 | ㉔ 18 ×12 | ㉕ 20 × 9 | ㉖ 12 ×14 | ㉗ 16 × 2 | ㉘ 11 ×15 | ㉙ 17 × 5 | ㉚ 14 ×11 |

| ㉛ 13 × 6 | ㉜ 19 ×18 | ㉝ 10 ×10 | ㉞ 16 ×16 | ㉟ 14 × 7 | ㊱ 11 ×17 | ㊲ 15 × 4 | ㊳ 17 ×19 | ㊴ 18 × 1 | ㊵ 20 ×13 |

Mixed Practice

① $\begin{array}{r} 16 \\ \times\ 4 \\ \hline \end{array}$
② $\begin{array}{r} 18 \\ \times 12 \\ \hline \end{array}$
③ $\begin{array}{r} 20 \\ \times 10 \\ \hline \end{array}$
④ $\begin{array}{r} 11 \\ \times 13 \\ \hline \end{array}$
⑤ $\begin{array}{r} 10 \\ \times\ 3 \\ \hline \end{array}$
⑥ $\begin{array}{r} 17 \\ \times 14 \\ \hline \end{array}$
⑦ $\begin{array}{r} 14 \\ \times\ 6 \\ \hline \end{array}$
⑧ $\begin{array}{r} 13 \\ \times 11 \\ \hline \end{array}$
⑨ $\begin{array}{r} 15 \\ \times\ 9 \\ \hline \end{array}$
⑩ $\begin{array}{r} 12 \\ \times 20 \\ \hline \end{array}$

⑪ $\begin{array}{r} 11 \\ \times\ 2 \\ \hline \end{array}$
⑫ $\begin{array}{r} 10 \\ \times 17 \\ \hline \end{array}$
⑬ $\begin{array}{r} 17 \\ \times\ 1 \\ \hline \end{array}$
⑭ $\begin{array}{r} 13 \\ \times 15 \\ \hline \end{array}$
⑮ $\begin{array}{r} 14 \\ \times\ 8 \\ \hline \end{array}$
⑯ $\begin{array}{r} 20 \\ \times 19 \\ \hline \end{array}$
⑰ $\begin{array}{r} 15 \\ \times\ 7 \\ \hline \end{array}$
⑱ $\begin{array}{r} 12 \\ \times 16 \\ \hline \end{array}$
⑲ $\begin{array}{r} 18 \\ \times\ 5 \\ \hline \end{array}$
⑳ $\begin{array}{r} 19 \\ \times 18 \\ \hline \end{array}$

㉑ $\begin{array}{r} 12 \\ \times\ 6 \\ \hline \end{array}$
㉒ $\begin{array}{r} 15 \\ \times 13 \\ \hline \end{array}$
㉓ $\begin{array}{r} 14 \\ \times\ 9 \\ \hline \end{array}$
㉔ $\begin{array}{r} 16 \\ \times 11 \\ \hline \end{array}$
㉕ $\begin{array}{r} 18 \\ \times\ 4 \\ \hline \end{array}$
㉖ $\begin{array}{r} 11 \\ \times 20 \\ \hline \end{array}$
㉗ $\begin{array}{r} 10 \\ \times 10 \\ \hline \end{array}$
㉘ $\begin{array}{r} 19 \\ \times 14 \\ \hline \end{array}$
㉙ $\begin{array}{r} 13 \\ \times\ 3 \\ \hline \end{array}$
㉚ $\begin{array}{r} 17 \\ \times 12 \\ \hline \end{array}$

㉛ $\begin{array}{r} 13 \\ \times\ 7 \\ \hline \end{array}$
㉜ $\begin{array}{r} 17 \\ \times 17 \\ \hline \end{array}$
㉝ $\begin{array}{r} 18 \\ \times\ 2 \\ \hline \end{array}$
㉞ $\begin{array}{r} 14 \\ \times 18 \\ \hline \end{array}$
㉟ $\begin{array}{r} 15 \\ \times\ 1 \\ \hline \end{array}$
㊱ $\begin{array}{r} 12 \\ \times 15 \\ \hline \end{array}$
㊲ $\begin{array}{r} 19 \\ \times\ 5 \\ \hline \end{array}$
㊳ $\begin{array}{r} 20 \\ \times 16 \\ \hline \end{array}$
㊴ $\begin{array}{r} 16 \\ \times\ 8 \\ \hline \end{array}$
㊵ $\begin{array}{r} 10 \\ \times 19 \\ \hline \end{array}$

Mixed Practice

① 14
×10

② 16
×20

③ 12
× 4

④ 10
×12

⑤ 20
× 6

⑥ 18
×13

⑦ 17
× 3

⑧ 11
×14

⑨ 19
× 9

⑩ 15
×11

⑪ 15
× 5

⑫ 20
×15

⑬ 11
× 8

⑭ 19
×17

⑮ 17
× 2

⑯ 13
×16

⑰ 16
× 1

⑱ 18
×18

⑲ 12
× 7

⑳ 14
×19

㉑ 10
× 4

㉒ 12
×14

㉓ 19
× 3

㉔ 20
×11

㉕ 16
× 9

㉖ 15
×12

㉗ 18
× 6

㉘ 14
×13

㉙ 17
×10

㉚ 13
×20

㉛ 19
× 7

㉜ 11
×17

㉝ 10
× 1

㉞ 18
×19

㉟ 13
× 8

㊱ 14
×16

㊲ 12
× 5

㊳ 17
×15

㊴ 20
× 2

㊵ 16
×18

Mixed Practice

① 12 ② 19 ③ 20 ④ 10 ⑤ 15 ⑥ 14 ⑦ 18 ⑧ 17 ⑨ 16 ⑩ 13
× 7   ×11   × 8   ×13   × 4   ×18   × 1   ×19   × 6   ×17

⑪ 10 ⑫ 15 ⑬ 14 ⑭ 17 ⑮ 18 ⑯ 20 ⑰ 16 ⑱ 13 ⑲ 19 ⑳ 11
×10   ×15   × 9   ×16   × 3   ×12   × 5   ×14   × 2   ×20

㉑ 13 ㉒ 16 ㉓ 18 ㉔ 12 ㉕ 19 ㉖ 10 ㉗ 15 ㉘ 11 ㉙ 17 ㉚ 14
× 1   ×13   × 6   ×19   × 7   ×17   × 8   ×18   × 4   ×11

㉛ 17 ㉜ 14 ㉝ 19 ㉞ 18 ㉟ 16 ㊱ 13 ㊲ 11 ㊳ 20 ㊴ 12 ㊵ 15
× 5   ×15   ×10   ×20   × 9   ×16   × 2   ×14   × 3   ×12

Mixed Practice

① 18   ② 12   ③ 13   ④ 15   ⑤ 20   ⑥ 19   ⑦ 14   ⑧ 10   ⑨ 11   ⑩ 16
  × 8   ×17   × 7   ×11   × 1   ×13   × 4   ×18   × 6   ×19
_____ _____ _____ _____ _____ _____ _____ _____ _____ _____

⑪ 16   ⑫ 20   ⑬ 10   ⑭ 11   ⑮ 14   ⑯ 17   ⑰ 12   ⑱ 19   ⑲ 13   ⑳ 18
  × 2   ×16   × 3   ×15   ×10   ×14   × 9   ×20   × 5   ×12
_____ _____ _____ _____ _____ _____ _____ _____ _____ _____

㉑ 15   ㉒ 13   ㉓ 11   ㉔ 20   ㉕ 12   ㉖ 16   ㉗ 19   ㉘ 18   ㉙ 14   ㉚ 17
  × 7   ×18   × 4   ×19   × 6   ×11   × 1   ×13   × 8   ×17
_____ _____ _____ _____ _____ _____ _____ _____ _____ _____

㉛ 11   ㉜ 10   ㉝ 15   ㉞ 19   ㉟ 17   ㊱ 18   ㊲ 13   ㊳ 14   ㊴ 20   ㊵ 12
  × 5   ×15   × 9   ×12   × 3   ×14   × 2   ×16   ×10   ×20
_____ _____ _____ _____ _____ _____ _____ _____ _____ _____

Mixed Practice

① 14
× 9

② 20
×15

③ 11
× 7

④ 12
×20

⑤ 15
× 3

⑥ 17
×19

⑦ 18
× 6

⑧ 10
×13

⑨ 13
× 1

⑩ 19
×16

⑪ 12
× 4

⑫ 15
×12

⑬ 17
× 2

⑭ 10
×11

⑮ 18
× 8

⑯ 11
×17

⑰ 13
× 5

⑱ 19
×14

⑲ 20
×10

⑳ 16
×18

㉑ 19
× 6

㉒ 13
×20

㉓ 18
× 1

㉔ 14
×13

㉕ 20
× 9

㉖ 12
×16

㉗ 15
× 7

㉘ 16
×19

㉙ 10
× 3

㉚ 17
×15

㉛ 10
× 5

㉜ 17
×12

㉝ 20
× 4

㉞ 18
×18

㉟ 13
× 2

㊱ 19
×11

㊲ 16
×10

㊳ 11
×14

㊴ 14
× 8

㊵ 15
×17

Mixed Practice

① 18 × 7 ② 14 ×16 ③ 19 × 9 ④ 15 ×15 ⑤ 11 × 6 ⑥ 20 ×20 ⑦ 17 × 3 ⑧ 12 ×19 ⑨ 16 × 1 ⑩ 13 ×13

⑪ 13 ×10 ⑫ 11 ×11 ⑬ 12 × 8 ⑭ 16 ×12 ⑮ 17 × 4 ⑯ 10 ×14 ⑰ 14 × 2 ⑱ 20 ×18 ⑲ 19 × 5 ⑳ 18 ×17

㉑ 15 × 9 ㉒ 19 ×19 ㉓ 16 × 3 ㉔ 11 ×13 ㉕ 14 × 1 ㉖ 13 ×15 ㉗ 20 × 6 ㉘ 18 ×20 ㉙ 17 × 7 ㉚ 10 ×16

㉛ 16 × 5 ㉜ 12 ×12 ㉝ 15 × 2 ㉞ 20 ×17 ㉟ 10 × 8 ㊱ 18 ×14 ㊲ 19 ×10 ㊳ 17 ×11 ㊴ 11 × 4 ㊵ 14 ×18

Mixed Practice

① 15   ② 16   ③ 12   ④ 14   ⑤ 18   ⑥ 20   ⑦ 10   ⑧ 13   ⑨ 19   ⑩ 17
× 6   ×13   ×10   ×14   × 8   ×19   × 4   ×11   × 2   ×18

⑪ 14   ⑫ 18   ⑬ 20   ⑭ 13   ⑮ 10   ⑯ 12   ⑰ 19   ⑱ 17   ⑲ 16   ⑳ 11
× 7   ×20   × 1   ×12   × 9   ×15   × 3   ×16   × 5   ×17

㉑ 17   ㉒ 19   ㉓ 10   ㉔ 15   ㉕ 16   ㉖ 14   ㉗ 18   ㉘ 11   ㉙ 13   ㉚ 20
× 4   ×14   × 2   ×11   × 6   ×18   ×10   ×19   × 8   ×13

㉛ 13   ㉜ 20   ㉝ 16   ㉞ 10   ㉟ 19   ㊱ 17   ㊲ 11   ㊳ 12   ㊴ 15   ㊵ 18
× 3   ×20   × 7   ×17   × 1   ×12   × 5   ×16   × 9   ×15

Mixed Practice

① 10
  ×10
___

② 15
  ×18
___

③ 17
  × 6
___

④ 18
  ×13
___

⑤ 12
  × 4
___

⑥ 16
  ×14
___

⑦ 20
  × 8
___

⑧ 14
  ×19
___

⑨ 11
  × 2
___

⑩ 19
  ×11
___

⑪ 19
  × 5
___

⑫ 12
  ×12
___

⑬ 14
  × 9
___

⑭ 11
  ×20
___

⑮ 20
  × 7
___

⑯ 13
  ×16
___

⑰ 15
  × 1
___

⑱ 16
  ×17
___

⑲ 17
  × 3
___

⑳ 10
  ×15
___

㉑ 18
  × 6
___

㉒ 17
  ×19
___

㉓ 11
  × 8
___

㉔ 12
  ×11
___

㉕ 15
  × 2
___

㉖ 19
  ×13
___

㉗ 16
  × 4
___

㉘ 10
  ×14
___

㉙ 20
  ×10
___

㉚ 13
  ×18
___

㉛ 11
  × 3
___

㉜ 14
  ×20
___

㉝ 18
  × 1
___

㉞ 16
  ×15
___

㉟ 13
  × 9
___

㊱ 10
  ×16
___

㊲ 17
  × 5
___

㊳ 20
  ×12
___

㊴ 12
  × 7
___

㊵ 15
  ×17
___

Two Digits Times Two Digits

① 10 ×15   ② 13 ×18   ③ 19 ×16   ④ 15 ×10   ⑤ 18 ×19   ⑥ 11 ×11   ⑦ 14 ×17   ⑧ 20 ×14   ⑨ 12 ×20   ⑩ 16 ×12

⑪ 14 ×20   ⑫ 17 ×11   ⑬ 20 ×15   ⑭ 10 ×18   ⑮ 19 ×19   ⑯ 15 ×12   ⑰ 12 ×16   ⑱ 13 ×13   ⑲ 18 ×10   ⑳ 11 ×17

㉑ 19 ×12   ㉒ 11 ×13   ㉓ 18 ×20   ㉔ 12 ×11   ㉕ 20 ×16   ㉖ 13 ×17   ㉗ 16 ×10   ㉘ 10 ×19   ㉙ 17 ×18   ㉚ 14 ×14

㉛ 12 ×19   ㉜ 20 ×10   ㉝ 13 ×12   ㉞ 16 ×15   ㉟ 11 ×18   ㊱ 10 ×14   ㊲ 17 ×17   ㊳ 19 ×11   ㊴ 15 ×20   ㊵ 18 ×13

Two Digits Times Two Digits

①
15
×13
___

②
18
×14
___

③
10
×10
___

④
14
×16
___

⑤
17
×12
___

⑥
16
×19
___

⑦
11
×20
___

⑧
12
×18
___

⑨
20
×11
___

⑩
19
×15
___

⑪
11
×12
___

⑫
16
×17
___

⑬
15
×11
___

⑭
19
×14
___

⑮
13
×10
___

⑯
14
×13
___

⑰
20
×19
___

⑱
18
×16
___

⑲
10
×15
___

⑳
17
×20
___

㉑
20
×18
___

㉒
19
×20
___

㉓
17
×16
___

㉔
18
×17
___

㉕
10
×12
___

㉖
15
×15
___

㉗
13
×14
___

㉘
16
×13
___

㉙
14
×11
___

㉚
12
×10
___

㉛
13
×11
___

㉜
14
×10
___

㉝
12
×13
___

㉞
11
×15
___

㉟
16
×16
___

㊱
17
×19
___

㊲
18
×12
___

㊳
15
×14
___

㊴
19
×18
___

㊵
20
×17
___

Two Digits Times Two Digits

① 18 ×13   ② 16 ×20   ③ 14 ×12   ④ 11 ×15   ⑤ 12 ×18   ⑥ 13 ×11   ⑦ 10 ×19   ⑧ 20 ×16   ⑨ 19 ×10   ⑩ 15 ×14

⑪ 10 ×10   ⑫ 17 ×11   ⑬ 20 ×13   ⑭ 18 ×20   ⑮ 14 ×18   ⑯ 11 ×14   ⑰ 19 ×12   ⑱ 16 ×17   ⑲ 12 ×15   ⑳ 13 ×19

㉑ 14 ×14   ㉒ 13 ×17   ㉓ 12 ×10   ㉔ 19 ×11   ㉕ 20 ×12   ㉖ 16 ×19   ㉗ 15 ×15   ㉘ 18 ×18   ㉙ 17 ×20   ㉚ 10 ×16

㉛ 19 ×18   ㉜ 20 ×15   ㉝ 16 ×14   ㉞ 15 ×13   ㉟ 13 ×20   ㊱ 18 ×16   ㊲ 17 ×19   ㊳ 14 ×11   ㊴ 11 ×10   ㊵ 12 ×17

Two Digits Times Two Digits

① 11 ×17   ② 12 ×16   ③ 18 ×15   ④ 10 ×12   ⑤ 17 ×14   ⑥ 15 ×18   ⑦ 13 ×10   ⑧ 19 ×20   ⑨ 20 ×11   ⑩ 14 ×13

⑪ 13 ×14   ⑫ 15 ×19   ⑬ 11 ×11   ⑭ 14 ×16   ⑮ 16 ×15   ⑯ 10 ×17   ⑰ 20 ×18   ⑱ 12 ×12   ⑲ 18 ×13   ⑳ 17 ×10

㉑ 20 ×20   ㉒ 14 ×10   ㉓ 17 ×12   ㉔ 12 ×19   ㉕ 18 ×14   ㉖ 11 ×13   ㉗ 16 ×16   ㉘ 15 ×17   ㉙ 10 ×11   ㉚ 19 ×15

㉛ 16 ×11   ㉜ 10 ×15   ㉝ 19 ×17   ㉞ 13 ×13   ㉟ 15 ×12   ㊱ 17 ×18   ㊲ 12 ×14   ㊳ 11 ×16   ㊴ 14 ×20   ㊵ 20 ×19

Two Digits Times Two Digits

①	②	③	④	⑤	⑥	⑦	⑧	⑨	⑩
13	12	15	14	19	20	11	17	18	10
×10	×17	×13	×14	×18	×15	×16	×12	×19	×11

⑪	⑫	⑬	⑭	⑮	⑯	⑰	⑱	⑲	⑳
11	16	17	13	15	14	18	12	19	20
×19	×15	×10	×17	×18	×11	×13	×20	×14	×16

㉑	㉒	㉓	㉔	㉕	㉖	㉗	㉘	㉙	㉚
15	20	19	18	17	12	10	13	16	11
×11	×20	×19	×15	×13	×16	×14	×18	×17	×12

㉛	㉜	㉝	㉞	㉟	㊱	㊲	㊳	㊴	㊵
18	17	12	10	20	13	16	15	14	19
×18	×14	×11	×10	×17	×12	×16	×15	×19	×20

Two Digits Times Two Digits

①	②	③	④	⑤	⑥	⑦	⑧	⑨	⑩
14	19	13	11	16	10	20	18	17	15
×20	×12	×14	×13	×11	×18	×19	×17	×15	×10

⑪	⑫	⑬	⑭	⑮	⑯	⑰	⑱	⑲	⑳
20	10	14	15	12	11	17	19	13	16
×11	×16	×15	×12	×14	×20	×18	×13	×10	×19

㉑	㉒	㉓	㉔	㉕	㉖	㉗	㉘	㉙	㉚
17	15	16	19	13	14	12	10	11	18
×17	×19	×13	×16	×11	×10	×12	×20	×15	×14

㉛	㉜	㉝	㉞	㉟	㊱	㊲	㊳	㊴	㊵
12	11	18	20	10	16	19	14	15	17
×15	×14	×20	×10	×13	×18	×11	×12	×17	×16

Two Digits Times Two Digits

① 12
×15

② 16
×18

③ 17
×10

④ 14
×17

⑤ 10
×12

⑥ 20
×19

⑦ 18
×14

⑧ 11
×11

⑨ 13
×20

⑩ 19
×13

⑪ 18
×20

⑫ 15
×19

⑬ 11
×15

⑭ 12
×18

⑮ 17
×12

⑯ 14
×13

⑰ 13
×10

⑱ 16
×16

⑲ 10
×17

⑳ 20
×14

㉑ 17
×13

㉒ 20
×16

㉓ 10
×20

㉔ 13
×19

㉕ 11
×10

㉖ 16
×14

㉗ 19
×17

㉘ 12
×12

㉙ 15
×18

㉚ 18
×11

㉛ 13
×12

㉜ 11
×17

㉝ 16
×13

㉞ 19
×15

㉟ 20
×18

㊱ 12
×11

㊲ 15
×14

㊳ 17
×19

㊴ 14
×20

㊵ 10
×16

Two Digits Times Two Digits

① 14 ×16 ② 10 ×11 ③ 12 ×17 ④ 18 ×10 ⑤ 15 ×13 ⑥ 19 ×12 ⑦ 20 ×20 ⑧ 13 ×18 ⑨ 11 ×19 ⑩ 17 ×15

⑪ 20 ×13 ⑫ 19 ×14 ⑬ 14 ×19 ⑭ 17 ×11 ⑮ 16 ×17 ⑯ 18 ×16 ⑰ 11 ×12 ⑱ 10 ×10 ⑲ 12 ×15 ⑳ 15 ×20

㉑ 11 ×18 ㉒ 17 ×20 ㉓ 15 ×10 ㉔ 10 ×14 ㉕ 12 ×13 ㉖ 14 ×15 ㉗ 16 ×11 ㉘ 19 ×16 ㉙ 18 ×19 ㉚ 13 ×17

㉛ 16 ×19 ㉜ 18 ×17 ㉝ 13 ×16 ㉞ 20 ×15 ㉟ 19 ×10 ㊱ 15 ×12 ㊲ 10 ×13 ㊳ 14 ×11 ㊴ 17 ×18 ㊵ 11 ×14

Two Digits Times Two Digits

①   ②   ③   ④   ⑤   ⑥   ⑦   ⑧   ⑨   ⑩

| 11 | 20 | 18 | 17 | 16 | 10 | 19 | 15 | 14 | 13 |
| ×16 | ×11 | ×19 | ×14 | ×18 | ×10 | ×15 | ×17 | ×12 | ×20 |

⑪   ⑫   ⑬   ⑭   ⑮   ⑯   ⑰   ⑱   ⑲   ⑳

| 19 | 12 | 15 | 11 | 18 | 17 | 14 | 20 | 16 | 10 |
| ×12 | ×10 | ×16 | ×11 | ×18 | ×20 | ×19 | ×13 | ×14 | ×15 |

㉑   ㉒   ㉓   ㉔   ㉕   ㉖   ㉗   ㉘   ㉙   ㉚

| 18 | 10 | 16 | 14 | 15 | 20 | 13 | 11 | 12 | 19 |
| ×20 | ×13 | ×12 | ×10 | ×19 | ×15 | ×14 | ×18 | ×11 | ×17 |

㉛   ㉜   ㉝   ㉞   ㉟   ㊱   ㊲   ㊳   ㊴   ㊵

| 14 | 15 | 20 | 13 | 10 | 11 | 12 | 18 | 17 | 16 |
| ×18 | ×14 | ×20 | ×16 | ×11 | ×17 | ×15 | ×10 | ×12 | ×13 |

Two Digits Times Two Digits

① 17 ×13

② 16 ×17

③ 11 ×14

④ 19 ×19

⑤ 12 ×20

⑥ 13 ×18

⑦ 10 ×12

⑧ 14 ×11

⑨ 15 ×10

⑩ 18 ×16

⑪ 10 ×20

⑫ 13 ×15

⑬ 17 ×10

⑭ 18 ×17

⑮ 20 ×14

⑯ 19 ×13

⑰ 15 ×18

⑱ 16 ×19

⑲ 11 ×16

⑳ 12 ×12

㉑ 15 ×11

㉒ 18 ×12

㉓ 12 ×19

㉔ 16 ×15

㉕ 11 ×20

㉖ 17 ×16

㉗ 20 ×17

㉘ 13 ×13

㉙ 19 ×10

㉚ 14 ×14

㉛ 20 ×10

㉜ 19 ×14

㉝ 14 ×13

㉞ 10 ×16

㉟ 13 ×19

㊱ 12 ×18

㊲ 16 ×20

㊳ 17 ×17

㊴ 18 ×11

㊵ 15 ×15

Two Digits Times Two Digits

① 19 ×12 ② 20 ×14 ③ 13 ×20 ④ 12 ×18 ⑤ 16 ×15 ⑥ 18 ×10 ⑦ 14 ×17 ⑧ 17 ×19 ⑨ 11 ×13 ⑩ 15 ×16

⑪ 14 ×13 ⑫ 10 ×10 ⑬ 17 ×12 ⑭ 19 ×14 ⑮ 13 ×15 ⑯ 12 ×16 ⑰ 11 ×20 ⑱ 20 ×11 ⑲ 16 ×18 ⑳ 18 ×17

㉑ 13 ×16 ㉒ 18 ×11 ㉓ 16 ×13 ㉔ 11 ×10 ㉕ 17 ×20 ㉖ 20 ×17 ㉗ 15 ×18 ㉘ 19 ×15 ㉙ 10 ×14 ㉚ 14 ×19

㉛ 11 ×15 ㉜ 17 ×18 ㉝ 20 ×16 ㉞ 15 ×12 ㉟ 18 ×14 ㊱ 19 ×19 ㊲ 10 ×17 ㊳ 13 ×10 ㊴ 12 ×13 ㊵ 16 ×11

Two Digits Times Two Digits

① 12
×11
___

② 16
×19
___

③ 19
×18
___

④ 14
×20
___

⑤ 10
×16
___

⑥ 15
×15
___

⑦ 18
×13
___

⑧ 11
×14
___

⑨ 17
×10
___

⑩ 13
×12
___

⑪ 18
×16
___

⑫ 15
×17
___

⑬ 12
×10
___

⑭ 13
×19
___

⑮ 20
×18
___

⑯ 14
×11
___

⑰ 17
×15
___

⑱ 16
×20
___

⑲ 19
×12
___

⑳ 10
×13
___

㉑ 17
×14
___

㉒ 13
×13
___

㉓ 10
×20
___

㉔ 16
×17
___

㉕ 19
×16
___

㉖ 12
×12
___

㉗ 20
×19
___

㉘ 15
×11
___

㉙ 14
×10
___

㉚ 11
×18
___

㉛ 20
×10
___

㉜ 14
×18
___

㉝ 11
×11
___

㉞ 18
×12
___

㉟ 15
×20
___

㊱ 10
×15
___

㊲ 16
×16
___

㊳ 12
×19
___

㊴ 13
×14
___

㊵ 17
×17
___

Two Digits Times Two Digits

① 15 ×18 ② 10 ×10 ③ 16 ×14 ④ 18 ×12 ⑤ 20 ×19 ⑥ 12 ×15 ⑦ 19 ×20 ⑧ 11 ×11 ⑨ 13 ×17 ⑩ 14 ×16

⑪ 19 ×17 ⑫ 17 ×15 ⑬ 11 ×18 ⑭ 15 ×10 ⑮ 16 ×19 ⑯ 18 ×16 ⑰ 13 ×14 ⑱ 10 ×13 ⑲ 20 ×12 ⑳ 12 ×20

㉑ 16 ×16 ㉒ 12 ×13 ㉓ 20 ×17 ㉔ 13 ×15 ㉕ 11 ×14 ㉖ 10 ×20 ㉗ 14 ×12 ㉘ 15 ×19 ㉙ 17 ×10 ㉚ 19 ×11

㉛ 13 ×19 ㉜ 11 ×12 ㉝ 10 ×16 ㉞ 14 ×18 ㉟ 12 ×10 ㊱ 15 ×11 ㊲ 17 ×20 ㊳ 16 ×15 ㊴ 18 ×17 ㊵ 20 ×13

Two Digits Times Two Digits

①
18
×13
___

②
20
×11
___

③
15
×12
___

④
19
×14
___

⑤
17
×16
___

⑥
14
×19
___

⑦
12
×17
___

⑧
13
×10
___

⑨
11
×15
___

⑩
16
×18
___

⑪
12
×16
___

⑫
14
×20
___

⑬
18
×15
___

⑭
16
×11
___

⑮
10
×12
___

⑯
19
×13
___

⑰
11
×19
___

⑱
20
×14
___

⑲
15
×18
___

⑳
17
×17
___

㉑
11
×10
___

㉒
16
×17
___

㉓
17
×14
___

㉔
20
×20
___

㉕
15
×16
___

㉖
18
×18
___

㉗
10
×11
___

㉘
14
×13
___

㉙
19
×15
___

㉚
13
×12
___

㉛
10
×15
___

㉜
19
×12
___

㉝
13
×13
___

㉞
12
×18
___

㉟
14
×14
___

㊱
17
×19
___

㊲
20
×16
___

㊳
18
×11
___

㊴
16
×10
___

㊵
11
×20
___

Two Digits Times Two Digits

①   ②   ③   ④   ⑤   ⑥   ⑦   ⑧   ⑨   ⑩

| 20 | 10 | 13 | 17 | 16 | 19 | 18 | 12 | 14 | 15 |
| ×18 | ×19 | ×15 | ×20 | ×12 | ×13 | ×10 | ×17 | ×11 | ×16 |

⑪   ⑫   ⑬   ⑭   ⑮   ⑯   ⑰   ⑱   ⑲   ⑳

| 18 | 11 | 12 | 20 | 13 | 17 | 14 | 10 | 16 | 19 |
| ×11 | ×13 | ×18 | ×19 | ×12 | ×16 | ×15 | ×14 | ×20 | ×10 |

㉑   ㉒   ㉓   ㉔   ㉕   ㉖   ㉗   ㉘   ㉙   ㉚

| 13 | 19 | 16 | 14 | 12 | 10 | 15 | 20 | 11 | 18 |
| ×16 | ×14 | ×11 | ×13 | ×15 | ×10 | ×20 | ×12 | ×19 | ×17 |

㉛   ㉜   ㉝   ㉞   ㉟   ㊱   ㊲   ㊳   ㊴   ㊵

| 14 | 12 | 10 | 15 | 19 | 20 | 11 | 13 | 17 | 16 |
| ×12 | ×20 | ×16 | ×18 | ×19 | ×17 | ×10 | ×13 | ×11 | ×14 |

Two Digits Times Two Digits

① 17   ② 16   ③ 20   ④ 18   ⑤ 11   ⑥ 15   ⑦ 19   ⑧ 14   ⑨ 12   ⑩ 13
  ×14   ×17   ×20   ×15   ×16   ×12   ×11   ×19   ×13   ×18
  ____   ____   ____   ____   ____   ____   ____   ____   ____   ____

⑪ 19   ⑫ 15   ⑬ 17   ⑭ 13   ⑮ 10   ⑯ 18   ⑰ 12   ⑱ 16   ⑲ 20   ⑳ 11
  ×16   ×10   ×13   ×17   ×20   ×14   ×12   ×15   ×18   ×11
  ____   ____   ____   ____   ____   ____   ____   ____   ____   ____

㉑ 12   ㉒ 13   ㉓ 11   ㉔ 16   ㉕ 20   ㉖ 17   ㉗ 10   ㉘ 15   ㉙ 18   ㉚ 14
  ×19   ×11   ×15   ×10   ×16   ×18   ×17   ×14   ×13   ×20
  ____   ____   ____   ____   ____   ____   ____   ____   ____   ____

㉛ 10   ㉜ 18   ㉝ 14   ㉞ 19   ㉟ 15   ㊱ 11   ㊲ 16   ㊳ 17   ㊴ 13   ㊵ 12
  ×13   ×20   ×14   ×18   ×15   ×12   ×16   ×17   ×19   ×10
  ____   ____   ____   ____   ____   ____   ____   ____   ____   ____

Two Digits Times Two Digits

① 19 ×12
② 18 ×10
③ 17 ×19
④ 14 ×13
⑤ 11 ×14
⑥ 20 ×11
⑦ 10 ×20
⑧ 12 ×15
⑨ 13 ×18
⑩ 16 ×16

⑪ 10 ×18
⑫ 15 ×11
⑬ 12 ×12
⑭ 19 ×10
⑮ 17 ×14
⑯ 14 ×16
⑰ 13 ×19
⑱ 18 ×17
⑲ 11 ×13
⑳ 20 ×20

㉑ 17 ×16
㉒ 20 ×17
㉓ 11 ×18
㉔ 13 ×11
㉕ 12 ×19
㉖ 18 ×20
㉗ 16 ×13
㉘ 19 ×14
㉙ 15 ×10
㉚ 10 ×15

㉛ 13 ×14
㉜ 12 ×13
㉝ 18 ×16
㉞ 16 ×12
㉟ 20 ×10
㊱ 19 ×15
㊲ 15 ×20
㊳ 17 ×11
㊴ 14 ×18
㊵ 11 ×17

Two Digits Times Two Digits

① 14 ×17   ② 11 ×15   ③ 19 ×13   ④ 10 ×19   ⑤ 15 ×16   ⑥ 16 ×14   ⑦ 20 ×18   ⑧ 13 ×10   ⑨ 12 ×11   ⑩ 17 ×12

⑪ 20 ×16   ⑫ 16 ×20   ⑬ 14 ×11   ⑭ 17 ×15   ⑮ 18 ×13   ⑯ 10 ×17   ⑰ 12 ×14   ⑱ 11 ×19   ⑲ 19 ×12   ⑳ 15 ×18

㉑ 12 ×10   ㉒ 17 ×18   ㉓ 15 ×19   ㉔ 11 ×20   ㉕ 19 ×16   ㉖ 14 ×12   ㉗ 18 ×15   ㉘ 16 ×17   ㉙ 10 ×11   ㉚ 13 ×13

㉛ 18 ×11   ㉜ 10 ×13   ㉝ 13 ×17   ㉞ 20 ×12   ㉟ 16 ×19   ㊱ 15 ×14   ㊲ 11 ×16   ㊳ 14 ×15   ㊴ 17 ×10   ㊵ 12 ×20

Two Digits Times Two Digits

① 14 ×14   ② 18 ×16   ③ 12 ×20   ④ 15 ×19   ⑤ 17 ×18   ⑥ 10 ×13   ⑦ 11 ×12   ⑧ 13 ×11   ⑨ 20 ×10   ⑩ 19 ×15

⑪ 11 ×10   ⑫ 16 ×13   ⑬ 13 ×14   ⑭ 14 ×16   ⑮ 12 ×18   ⑯ 15 ×15   ⑰ 20 ×20   ⑱ 18 ×17   ⑲ 17 ×19   ⑳ 10 ×12

㉑ 12 ×15   ㉒ 10 ×17   ㉓ 17 ×10   ㉔ 20 ×13   ㉕ 13 ×20   ㉖ 18 ×12   ㉗ 19 ×19   ㉘ 14 ×18   ㉙ 16 ×16   ㉚ 11 ×11

㉛ 20 ×18   ㉜ 13 ×19   ㉝ 18 ×15   ㉞ 19 ×14   ㉟ 10 ×16   ㊱ 14 ×11   ㊲ 16 ×12   ㊳ 12 ×13   ㊴ 15 ×10   ㊵ 17 ×17

Two Digits Times Two Digits

①	②	③	④	⑤	⑥	⑦	⑧	⑨	⑩
15	17	14	11	16	19	10	20	13	12
×17	×11	×19	×20	×15	×18	×10	×16	×13	×14

⑪	⑫	⑬	⑭	⑮	⑯	⑰	⑱	⑲	⑳
10	19	15	12	18	11	13	17	14	16
×15	×12	×13	×11	×19	×17	×18	×20	×14	×10

㉑	㉒	㉓	㉔	㉕	㉖	㉗	㉘	㉙	㉚
13	12	16	17	14	15	18	19	11	20
×16	×10	×20	×12	×15	×14	×11	×17	×13	×19

㉛	㉜	㉝	㉞	㉟	㊱	㊲	㊳	㊴	㊵
18	11	20	10	19	16	17	15	12	13
×13	×19	×17	×14	×20	×18	×15	×11	×16	×12

Review 1-9

① 3 × 4   ② 7 × 5   ③ 2 × 3   ④ 5 × 6   ⑤ 6 × 4   ⑥ 4 × 8   ⑦ 9 × 9   ⑧ 8 × 7   ⑨ 3 × 2   ⑩ 1 × 5

⑪ 6 × 2   ⑫ 1 × 3   ⑬ 8 × 9   ⑭ 9 × 7   ⑮ 3 × 8   ⑯ 2 × 7   ⑰ 5 × 9   ⑱ 4 × 2   ⑲ 7 × 3   ⑳ 8 × 6

㉑ 9 × 4   ㉒ 5 × 1   ㉓ 3 × 6   ㉔ 4 × 9   ㉕ 7 × 9   ㉖ 8 × 4   ㉗ 1 × 8   ㉘ 2 × 8   ㉙ 5 × 5   ㉚ 6 × 1

㉛ 8 × 5   ㉜ 4 × 4   ㉝ 7 × 7   ㉞ 2 × 1   ㉟ 5 × 2   ㊱ 6 × 6   ㊲ 3 × 1   ㊳ 9 × 5   ㊴ 1 × 7   ㊵ 4 × 3

Review 1-9

① 2 × 2

② 3 × 7

③ 9 × 6

④ 6 × 9

⑤ 4 × 5

⑥ 1 × 1

⑦ 7 × 1

⑧ 5 × 3

⑨ 8 × 8

⑩ 2 × 4

⑪ 7 × 6

⑫ 2 × 5

⑬ 5 × 8

⑭ 1 × 2

⑮ 8 × 1

⑯ 9 × 8

⑰ 4 × 6

⑱ 6 × 5

⑲ 9 × 3

⑳ 3 × 9

㉑ 1 × 4

㉒ 8 × 2

㉓ 4 × 7

㉔ 3 × 3

㉕ 9 × 1

㉖ 7 × 2

㉗ 6 × 7

㉘ 1 × 6

㉙ 2 × 9

㉚ 5 × 4

㉛ 5 × 7

㉜ 6 × 3

㉝ 1 × 9

㉞ 7 × 8

㉟ 2 × 6

㊱ 3 × 5

㊲ 8 × 3

㊳ 7 × 4

㊴ 4 × 1

㊵ 9 × 2

________________ (name) ________ (date) ________ (time) _____ / 40 (score)

Review 1-9

① 6×2	② 4×9	③ 9×4	④ 7×5	⑤ 2×2	⑥ 1×7	⑦ 3×3	⑧ 8×1	⑨ 6×8	⑩ 5×9
⑪ 2×8	⑫ 5×4	⑬ 8×3	⑭ 3×1	⑮ 6×7	⑯ 9×1	⑰ 7×3	⑱ 1×8	⑲ 4×4	⑳ 8×5
㉑ 3×2	㉒ 7×6	㉓ 6×5	㉔ 1×3	㉕ 4×3	㉖ 8×2	㉗ 5×7	㉘ 9×7	㉙ 7×9	㉚ 2×6
㉛ 8×9	㉜ 1×2	㉝ 4×1	㉞ 9×6	㉟ 7×8	㊱ 2×5	㊲ 6×6	㊳ 3×9	㊴ 5×1	㊵ 1×4

Review 1-9

① 9 × 8
② 6 × 1
③ 3 × 5
④ 2 × 3
⑤ 1 × 9
⑥ 5 × 6
⑦ 4 × 6
⑧ 7 × 4
⑨ 8 × 7
⑩ 9 × 2

⑪ 4 × 5
⑫ 9 × 9
⑬ 7 × 7
⑭ 5 × 8
⑮ 8 × 6
⑯ 3 × 7
⑰ 1 × 5
⑱ 2 × 9
⑲ 3 × 4
⑳ 6 × 3

㉑ 5 × 2
㉒ 8 × 8
㉓ 1 × 1
㉔ 6 × 4
㉕ 3 × 6
㉖ 4 × 8
㉗ 2 × 1
㉘ 5 × 5
㉙ 9 × 3
㉚ 7 × 2

㉛ 7 × 1
㉜ 2 × 4
㉝ 5 × 3
㉞ 4 × 7
㉟ 9 × 5
㊱ 6 × 9
㊲ 8 × 4
㊳ 4 × 2
㊴ 1 × 6
㊵ 3 × 8

Review 1-9

①   ②   ③   ④   ⑤   ⑥   ⑦   ⑧   ⑨   ⑩
    1   9   4   7   5   3   2   8   1   6
× 4   × 8   × 9   × 1   × 4   × 5   × 3   × 2   × 6   × 8
_____   _____   _____   _____   _____   _____   _____   _____   _____   _____

⑪   ⑫   ⑬   ⑭   ⑮   ⑯   ⑰   ⑱   ⑲   ⑳
    5   6   8   2   1   4   7   3   9   8
× 6   × 9   × 3   × 2   × 5   × 2   × 3   × 6   × 9   × 1
_____   _____   _____   _____   _____   _____   _____   _____   _____   _____

㉑   ㉒   ㉓   ㉔   ㉕   ㉖   ㉗   ㉘   ㉙   ㉚
    2   7   1   3   9   8   6   4   7   5
× 4   × 7   × 1   × 3   × 3   × 4   × 5   × 5   × 8   × 7
_____   _____   _____   _____   _____   _____   _____   _____   _____   _____

㉛   ㉜   ㉝   ㉞   ㉟   ㊱   ㊲   ㊳   ㊴   ㊵
    8   3   9   4   7   5   1   2   6   3
× 8   × 4   × 2   × 7   × 6   × 1   × 7   × 8   × 2   × 9
_____   _____   _____   _____   _____   _____   _____   _____   _____   _____

Review 1-9

① 4 × 6

② 1 × 2

③ 2 × 1

④ 5 × 3

⑤ 3 × 8

⑥ 6 × 7

⑦ 9 × 7

⑧ 7 × 9

⑨ 8 × 5

⑩ 4 × 4

⑪ 9 × 1

⑫ 4 × 8

⑬ 7 × 5

⑭ 6 × 6

⑮ 8 × 7

⑯ 2 × 5

⑰ 3 × 1

⑱ 5 × 8

⑲ 2 × 9

⑳ 1 × 3

㉑ 6 × 4

㉒ 8 × 6

㉓ 3 × 2

㉔ 1 × 9

㉕ 2 × 7

㉖ 9 × 6

㉗ 5 × 2

㉘ 6 × 1

㉙ 4 × 3

㉚ 7 × 4

㉛ 7 × 2

㉜ 5 × 9

㉝ 6 × 3

㉞ 9 × 5

㉟ 4 × 1

㊱ 1 × 8

㊲ 8 × 9

㊳ 9 × 4

㊴ 3 × 7

㊵ 2 × 6

Review 1-9

① 8 × 4

② 9 × 3

③ 2 × 7

④ 7 × 8

⑤ 1 × 4

⑥ 4 × 5

⑦ 5 × 6

⑧ 3 × 1

⑨ 8 × 9

⑩ 6 × 3

⑪ 1 × 9

⑫ 6 × 7

⑬ 3 × 6

⑭ 5 × 1

⑮ 8 × 5

⑯ 2 × 1

⑰ 7 × 6

⑱ 4 × 9

⑲ 9 × 7

⑳ 3 × 8

㉑ 5 × 4

㉒ 7 × 2

㉓ 8 × 8

㉔ 4 × 6

㉕ 9 × 6

㉖ 3 × 4

㉗ 6 × 5

㉘ 2 × 5

㉙ 7 × 3

㉚ 1 × 2

㉛ 3 × 3

㉜ 4 × 4

㉝ 9 × 1

㉞ 2 × 2

㉟ 7 × 9

㊱ 1 × 8

㊲ 8 × 2

㊳ 5 × 3

㊴ 6 × 1

㊵ 4 × 7

Review 1-9

① 2   ② 8   ③ 5   ④ 1   ⑤ 4   ⑥ 6   ⑦ 9   ⑧ 7   ⑨ 3   ⑩ 2
× 9   × 1   × 8   × 6   × 3   × 2   × 2   × 7   × 5   × 4

⑪ 9   ⑫ 2   ⑬ 7   ⑭ 6   ⑮ 3   ⑯ 5   ⑰ 4   ⑱ 1   ⑲ 5   ⑳ 8
× 8   × 3   × 5   × 9   × 2   × 5   × 8   × 3   × 7   × 6

㉑ 6   ㉒ 3   ㉓ 4   ㉔ 8   ㉕ 5   ㉖ 9   ㉗ 1   ㉘ 6   ㉙ 2   ㉚ 7
× 4   × 9   × 1   × 7   × 2   × 9   × 1   × 8   × 6   × 4

㉛ 7   ㉜ 1   ㉝ 6   ㉞ 9   ㉟ 2   ㊱ 8   ㊲ 3   ㊳ 9   ㊴ 4   ㊵ 5
× 1   × 7   × 6   × 5   × 8   × 3   × 7   × 4   × 2   × 9

Review 1-9

① 　2　　② 　7　　③ 　6　　④ 　5　　⑤ 　3　　⑥ 　9　　⑦ 　4　　⑧ 　8　　⑨ 　2　　⑩ 　1
× 1　　× 2　　× 8　　× 4　　× 1　　× 7　　× 5　　× 9　　× 3　　× 2

⑪ 　3　　⑫ 　1　　⑬ 　8　　⑭ 　4　　⑮ 　2　　⑯ 　6　　⑰ 　5　　⑱ 　9　　⑲ 　7　　⑳ 　8
× 3　　× 8　　× 5　　× 9　　× 7　　× 9　　× 5　　× 3　　× 8　　× 4

㉑ 　4　　㉒ 　5　　㉓ 　2　　㉔ 　9　　㉕ 　7　　㉖ 　8　　㉗ 　1　　㉘ 　6　　㉙ 　5　　㉚ 　3
× 1　　× 6　　× 4　　× 5　　× 5　　× 1　　× 7　　× 7　　× 2　　× 6

㉛ 　8　　㉜ 　9　　㉝ 　7　　㉞ 　6　　㉟ 　5　　㊱ 　3　　㊲ 　2　　㊳ 　4　　㊴ 　1　　㊵ 　9
× 2　　× 1　　× 9　　× 6　　× 3　　× 4　　× 6　　× 2　　× 9　　× 8

ANSWER KEY

Page 12
①80 ②170 ③50 ④200 ⑤60 ⑥160 ⑦10 ⑧110 ⑨100 ⑩150
⑪130 ⑫30 ⑬120 ⑭40 ⑮190 ⑯90 ⑰140 ⑱20 ⑲180 ⑳70
㉑110 ㉒20 ㉓140 ㉔100 ㉕170 ㉖70 ㉗120 ㉘80 ㉙200 ㉚30
㉛10 ㉜150 ㉝40 ㉞160 ㉟60 ㊱190 ㊲90 ㊳130 ㊴50 ㊵180

Page 13
①170 ②60 ③190 ④80 ⑤200 ⑥70 ⑦130 ⑧20 ⑨160 ⑩100
⑪140 ⑫40 ⑬150 ⑭10 ⑮160 ⑯50 ⑰120 ⑱60 ⑲170 ⑳30
㉑70 ㉒110 ㉓100 ㉔130 ㉕90 ㉖180 ㉗20 ㉘200 ㉙80 ㉚190
㉛60 ㉜190 ㉝30 ㉞200 ㉟20 ㊱150 ㊲10 ㊳160 ㊴100 ㊵140

Page 14
①55 ②165 ③99 ④132 ⑤110 ⑥154 ⑦77 ⑧143 ⑨22 ⑩121
⑪187 ⑫11 ⑬209 ⑭33 ⑮220 ⑯66 ⑰176 ⑱44 ⑲198 ⑳88
㉑143 ㉒44 ㉓176 ㉔22 ㉕165 ㉖88 ㉗209 ㉘55 ㉙132 ㉚11
㉛77 ㉜121 ㉝33 ㉞154 ㉟110 ㊱220 ㊲66 ㊳187 ㊴99 ㊵198

Page 15
①165 ②110 ③220 ④55 ⑤132 ⑥88 ⑦187 ⑧44 ⑨154 ⑩22
⑪176 ⑫33 ⑬121 ⑭77 ⑮154 ⑯99 ⑰209 ⑱110 ⑲165 ⑳11
㉑88 ㉒143 ㉓22 ㉔187 ㉕66 ㉖198 ㉗44 ㉘132 ㉙55 ㉚220
㉛110 ㉜220 ㉝11 ㉞132 ㉟44 ㊱121 ㊲77 ㊳154 ㊴22 ㊵176

Page 16
①72 ②228 ③96 ④156 ⑤36 ⑥240 ⑦60 ⑧168 ⑨48 ⑩180
⑪144 ⑫24 ⑬132 ⑭84 ⑮216 ⑯12 ⑰192 ⑱108 ⑲204 ⑳120
㉑168 ㉒108 ㉓192 ㉔48 ㉕228 ㉖120 ㉗132 ㉘72 ㉙156 ㉚24
㉛60 ㉜180 ㉝84 ㉞240 ㉟36 ㊱216 ㊲12 ㊳144 ㊴96 ㊵204

Page 17

① 228 ② 36   ③ 216 ④ 72   ⑤ 156 ⑥ 120 ⑦ 144 ⑧ 108 ⑨ 240 ⑩ 48
⑪ 192 ⑫ 84   ⑬ 180 ⑭ 60   ⑮ 240 ⑯ 96   ⑰ 132 ⑱ 36   ⑲ 228 ⑳ 24
㉑ 120 ㉒ 168 ㉓ 48   ㉔ 144 ㉕ 12   ㉖ 204 ㉗ 108 ㉘ 156 ㉙ 72   ㉚ 216
㉛ 36   ㉜ 216 ㉝ 24   ㉞ 156 ㉟ 108 ㊱ 180 ㊲ 60   ㊳ 240 ㊴ 48   ㊵ 192

Page 18

① 52   ② 234 ③ 130 ④ 182 ⑤ 78   ⑥ 260 ⑦ 91   ⑧ 195 ⑨ 104 ⑩ 208
⑪ 221 ⑫ 13   ⑬ 156 ⑭ 26   ⑮ 247 ⑯ 65   ⑰ 143 ⑱ 39   ⑲ 169 ⑳ 117
㉑ 195 ㉒ 39   ㉓ 143 ㉔ 104 ㉕ 234 ㉖ 117 ㉗ 156 ㉘ 52   ㉙ 182 ㉚ 13
㉛ 91   ㉜ 208 ㉝ 26   ㉞ 260 ㉟ 78   ㊱ 247 ㊲ 65   ㊳ 221 ㊴ 130 ㊵ 169

Page 19

① 234 ② 78   ③ 247 ④ 52   ⑤ 182 ⑥ 117 ⑦ 221 ⑧ 39   ⑨ 260 ⑩ 104
⑪ 143 ⑫ 26   ⑬ 208 ⑭ 91   ⑮ 260 ⑯ 130 ⑰ 156 ⑱ 78   ⑲ 234 ⑳ 13
㉑ 117 ㉒ 195 ㉓ 104 ㉔ 221 ㉕ 65   ㉖ 169 ㉗ 39   ㉘ 182 ㉙ 52   ㉚ 247
㉛ 78   ㉜ 247 ㉝ 13   ㉞ 182 ㉟ 39   ㊱ 208 ㊲ 91   ㊳ 260 ㊴ 104 ㊵ 143

Page 20

① 14   ② 168 ③ 42   ④ 252 ⑤ 28   ⑥ 266 ⑦ 70   ⑧ 224 ⑨ 126 ⑩ 182
⑪ 196 ⑫ 112 ⑬ 238 ⑭ 140 ⑮ 210 ⑯ 84   ⑰ 154 ⑱ 56   ⑲ 280 ⑳ 98
㉑ 224 ㉒ 56   ㉓ 154 ㉔ 126 ㉕ 168 ㉖ 98   ㉗ 238 ㉘ 14   ㉙ 252 ㉚ 112
㉛ 70   ㉜ 182 ㉝ 140 ㉞ 266 ㉟ 28   ㊱ 210 ㊲ 84   ㊳ 196 ㊴ 42   ㊵ 280

Page 21

① 168 ② 28   ③ 210 ④ 14   ⑤ 252 ⑥ 98   ⑦ 196 ⑧ 56   ⑨ 266 ⑩ 126
⑪ 154 ⑫ 140 ⑬ 182 ⑭ 70   ⑮ 266 ⑯ 42   ⑰ 238 ⑱ 28   ⑲ 168 ⑳ 112
㉑ 98   ㉒ 224 ㉓ 126 ㉔ 196 ㉕ 84   ㉖ 280 ㉗ 56   ㉘ 252 ㉙ 14   ㉚ 210
㉛ 28   ㉜ 210 ㉝ 112 ㉞ 252 ㉟ 56   ㊱ 182 ㊲ 70   ㊳ 266 ㊴ 126 ㊵ 154

Page 22

① 135 ② 225 ③ 45   ④ 210 ⑤ 30   ⑥ 285 ⑦ 15   ⑧ 255 ⑨ 90   ⑩ 195
⑪ 180 ⑫ 150 ⑬ 240 ⑭ 120 ⑮ 270 ⑯ 105 ⑰ 165 ⑱ 75   ⑲ 300 ⑳ 60
㉑ 255 ㉒ 75   ㉓ 165 ㉔ 90   ㉕ 225 ㉖ 60   ㉗ 240 ㉘ 135 ㉙ 210 ㉚ 150
㉛ 15   ㉜ 195 ㉝ 120 ㉞ 285 ㉟ 30   ㊱ 270 ㊲ 105 ㊳ 180 ㊴ 45   ㊵ 300

Page 23

① 225 ② 30 ③ 270 ④ 135 ⑤ 210 ⑥ 60 ⑦ 180 ⑧ 75 ⑨ 285 ⑩ 90
⑪ 165 ⑫ 120 ⑬ 195 ⑭ 15 ⑮ 285 ⑯ 45 ⑰ 240 ⑱ 30 ⑲ 225 ⑳ 150
㉑ 60 ㉒ 255 ㉓ 90 ㉔ 180 ㉕ 105 ㉖ 300 ㉗ 75 ㉘ 210 ㉙ 135 ㉚ 270
㉛ 30 ㉜ 270 ㉝ 150 ㉞ 210 ㉟ 75 ㊱ 195 ㊲ 15 ㊳ 285 ㊴ 90 ㊵ 165

Page 24

① 32 ② 288 ③ 96 ④ 176 ⑤ 16 ⑥ 256 ⑦ 112 ⑧ 272 ⑨ 64 ⑩ 240
⑪ 320 ⑫ 160 ⑬ 304 ⑭ 128 ⑮ 192 ⑯ 48 ⑰ 208 ⑱ 80 ⑲ 224 ⑳ 144
㉑ 272 ㉒ 80 ㉓ 208 ㉔ 64 ㉕ 288 ㉖ 144 ㉗ 304 ㉘ 32 ㉙ 176 ㉚ 160
㉛ 112 ㉜ 240 ㉝ 128 ㉞ 256 ㉟ 16 ㊱ 192 ㊲ 48 ㊳ 320 ㊴ 96 ㊵ 224

Page 25

① 288 ② 16 ③ 192 ④ 32 ⑤ 176 ⑥ 144 ⑦ 320 ⑧ 80 ⑨ 256 ⑩ 64
⑪ 208 ⑫ 128 ⑬ 240 ⑭ 112 ⑮ 256 ⑯ 96 ⑰ 304 ⑱ 16 ⑲ 288 ⑳ 160
㉑ 144 ㉒ 272 ㉓ 64 ㉔ 320 ㉕ 48 ㉖ 224 ㉗ 80 ㉘ 176 ㉙ 32 ㉚ 192
㉛ 16 ㉜ 192 ㉝ 160 ㉞ 176 ㉟ 80 ㊱ 240 ㊲ 112 ㊳ 256 ㊴ 64 ㊵ 208

Page 26

① 170 ② 187 ③ 85 ④ 323 ⑤ 102 ⑥ 204 ⑦ 17 ⑧ 272 ⑨ 34 ⑩ 306
⑪ 255 ⑫ 136 ⑬ 340 ⑭ 68 ⑮ 238 ⑯ 153 ⑰ 221 ⑱ 119 ⑲ 289 ⑳ 51
㉑ 272 ㉒ 119 ㉓ 221 ㉔ 34 ㉕ 187 ㉖ 51 ㉗ 340 ㉘ 170 ㉙ 323 ㉚ 136
㉛ 17 ㉜ 306 ㉝ 68 ㉞ 204 ㉟ 102 ㊱ 238 ㊲ 153 ㊳ 255 ㊴ 85 ㊵ 289

Page 27

① 187 ② 102 ③ 238 ④ 170 ⑤ 323 ⑥ 51 ⑦ 255 ⑧ 119 ⑨ 204 ⑩ 34
⑪ 221 ⑫ 68 ⑬ 306 ⑭ 17 ⑮ 204 ⑯ 85 ⑰ 340 ⑱ 102 ⑲ 187 ⑳ 136
㉑ 51 ㉒ 272 ㉓ 34 ㉔ 255 ㉕ 153 ㉖ 289 ㉗ 119 ㉘ 323 ㉙ 170 ㉚ 238
㉛ 102 ㉜ 238 ㉝ 136 ㉞ 323 ㉟ 119 ㊱ 306 ㊲ 17 ㊳ 204 ㊴ 34 ㊵ 221

Page 28

① 126 ② 198 ③ 18 ④ 216 ⑤ 180 ⑥ 360 ⑦ 108 ⑧ 270 ⑨ 144 ⑩ 252
⑪ 324 ⑫ 54 ⑬ 288 ⑭ 36 ⑮ 234 ⑯ 72 ⑰ 306 ⑱ 90 ⑲ 342 ⑳ 162
㉑ 270 ㉒ 90 ㉓ 306 ㉔ 144 ㉕ 198 ㉖ 162 ㉗ 288 ㉘ 126 ㉙ 216 ㉚ 54
㉛ 108 ㉜ 252 ㉝ 36 ㉞ 360 ㉟ 180 ㊱ 234 ㊲ 72 ㊳ 324 ㊴ 18 ㊵ 342

Page 29

① 198 ② 180 ③ 234 ④ 126 ⑤ 216 ⑥ 162 ⑦ 324 ⑧ 90 ⑨ 360 ⑩ 144
⑪ 306 ⑫ 36 ⑬ 252 ⑭ 108 ⑮ 360 ⑯ 18 ⑰ 288 ⑱ 180 ⑲ 198 ⑳ 54
㉑ 162 ㉒ 270 ㉓ 144 ㉔ 324 ㉕ 72 ㉖ 342 ㉗ 90 ㉘ 216 ㉙ 126 ㉚ 234
㉛ 180 ㉜ 234 ㉝ 54 ㉞ 216 ㉟ 90 ㊱ 252 ㊲ 108 ㊳ 360 ㊴ 144 ㊵ 306

Page 30

① 57 ② 380 ③ 171 ④ 323 ⑤ 38 ⑥ 247 ⑦ 76 ⑧ 304 ⑨ 133 ⑩ 285
⑪ 228 ⑫ 190 ⑬ 342 ⑭ 152 ⑮ 209 ⑯ 114 ⑰ 266 ⑱ 19 ⑲ 361 ⑳ 95
㉑ 304 ㉒ 19 ㉓ 266 ㉔ 133 ㉕ 380 ㉖ 95 ㉗ 342 ㉘ 57 ㉙ 323 ㉚ 190
㉛ 76 ㉜ 285 ㉝ 152 ㉞ 247 ㉟ 38 ㊱ 209 ㊲ 114 ㊳ 228 ㊴ 171 ㊵ 361

Page 31

① 380 ② 38 ③ 209 ④ 57 ⑤ 323 ⑥ 95 ⑦ 228 ⑧ 19 ⑨ 247 ⑩ 133
⑪ 266 ⑫ 152 ⑬ 285 ⑭ 76 ⑮ 247 ⑯ 171 ⑰ 342 ⑱ 38 ⑲ 380 ⑳ 190
㉑ 95 ㉒ 304 ㉓ 133 ㉔ 228 ㉕ 114 ㉖ 361 ㉗ 19 ㉘ 323 ㉙ 57 ㉚ 209
㉛ 38 ㉜ 209 ㉝ 190 ㉞ 323 ㉟ 19 ㊱ 285 ㊲ 76 ㊳ 247 ㊴ 133 ㊵ 266

Page 32

① 120 ② 380 ③ 40 ④ 260 ⑤ 160 ⑥ 300 ⑦ 100 ⑧ 360 ⑨ 20 ⑩ 400
⑪ 320 ⑫ 200 ⑬ 280 ⑭ 80 ⑮ 340 ⑯ 60 ⑰ 240 ⑱ 180 ⑲ 220 ⑳ 140
㉑ 360 ㉒ 180 ㉓ 240 ㉔ 20 ㉕ 380 ㉖ 140 ㉗ 280 ㉘ 120 ㉙ 260 ㉚ 200
㉛ 100 ㉜ 400 ㉝ 80 ㉞ 300 ㉟ 160 ㊱ 340 ㊲ 60 ㊳ 320 ㊴ 40 ㊵ 220

Page 33

① 380 ② 160 ③ 340 ④ 120 ⑤ 260 ⑥ 140 ⑦ 320 ⑧ 180 ⑨ 300 ⑩ 20
⑪ 240 ⑫ 80 ⑬ 400 ⑭ 100 ⑮ 300 ⑯ 40 ⑰ 280 ⑱ 160 ⑲ 380 ⑳ 200
㉑ 140 ㉒ 360 ㉓ 20 ㉔ 320 ㉕ 60 ㉖ 220 ㉗ 180 ㉘ 260 ㉙ 120 ㉚ 340
㉛ 160 ㉜ 340 ㉝ 200 ㉞ 260 ㉟ 180 ㊱ 400 ㊲ 100 ㊳ 300 ㊴ 20 ㊵ 240

Page 34

① 128 ② 17 ③ 120 ④ 75 ⑤ 126 ⑥ 48 ⑦ 28 ⑧ 152 ⑨ 117 ⑩ 33
⑪ 108 ⑫ 84 ⑬ 22 ⑭ 95 ⑮ 13 ⑯ 50 ⑰ 80 ⑱ 48 ⑲ 144 ⑳ 119
㉑ 180 ㉒ 30 ㉓ 30 ㉔ 16 ㉕ 57 ㉖ 66 ㉗ 90 ㉘ 136 ㉙ 56 ㉚ 91
㉛ 44 ㉜ 54 ㉝ 85 ㉞ 20 ㉟ 126 ㊱ 104 ㊲ 90 ㊳ 32 ㊴ 20 ㊵ 84

Page 35

① 14　② 60　③ 65　④ 36　⑤ 96　⑥ 48　⑦ 114　⑧ 140　⑨ 99　⑩ 60

⑪ 96　⑫ 160　⑬ 60　⑭ 11　⑮ 135　⑯ 76　⑰ 39　⑱ 40　⑲ 34　⑳ 98

㉑ 26　㉒ 171　㉓ 15　㉔ 90　㉕ 102　㉖ 112　㉗ 36　㉘ 112　㉙ 100　㉚ 72

㉛ 68　㉜ 24　㉝ 70　㉞ 78　㉟ 88　㊱ 60　㊲ 19　㊳ 162　㊴ 105　㊵ 10

Page 36

① 44　② 70　③ 36　④ 120　⑤ 153　⑥ 70　⑦ 57　⑧ 52　⑨ 120　⑩ 12

⑪ 84　⑫ 38　⑬ 36　⑭ 104　⑮ 140　⑯ 128　⑰ 90　⑱ 11　⑲ 68　⑳ 90

㉑ 108　㉒ 45　㉓ 16　㉔ 77　㉕ 13　㉖ 24　㉗ 136　㉘ 40　㉙ 95　㉚ 180

㉛ 60　㉜ 17　㉝ 80　㉞ 126　㉟ 114　㊱ 80　㊲ 30　㊳ 33　㊴ 48　㊵ 126

Page 37

① 133　② 32　③ 160　④ 51　⑤ 56　⑥ 11　⑦ 26　⑧ 162　⑨ 72　⑩ 75

⑪ 22　⑫ 72　⑬ 112　⑭ 84　⑮ 90　⑯ 65　⑰ 20　⑱ 80　⑲ 30　⑳ 171

㉑ 60　㉒ 78　㉓ 105　㉔ 96　㉕ 20　㉖ 76　㉗ 14　㉘ 99　㉙ 144　㉚ 85

㉛ 50　㉜ 42　㉝ 152　㉞ 40　㉟ 48　㊱ 18　㊲ 91　㊳ 102　㊴ 135　㊵ 112

Page 38

① 72　② 90　③ 22　④ 19　⑤ 91　⑥ 70　⑦ 96　⑧ 40　⑨ 60　⑩ 153

⑪ 42　⑫ 24　⑬ 136　⑭ 10　⑮ 120　⑯ 16　⑰ 55　⑱ 162　⑲ 52　⑳ 105

㉑ 33　㉒ 152　㉓ 144　㉔ 108　㉕ 90　㉖ 34　㉗ 13　㉘ 60　㉙ 60　㉚ 140

㉛ 85　㉜ 117　㉝ 15　㉞ 66　㉟ 36　㊱ 80　㊲ 38　㊳ 144　㊴ 128　㊵ 98

Page 39

① 72　② 32　③ 20　④ 104　⑤ 56　⑥ 162　⑦ 20　⑧ 77　⑨ 51　⑩ 95

⑪ 36　⑫ 44　⑬ 14　⑭ 102　⑮ 57　⑯ 50　⑰ 180　⑱ 80　⑲ 120　⑳ 84

㉑ 160　㉒ 30　㉓ 114　㉔ 48　㉕ 30　㉖ 48　㉗ 126　㉘ 126　㉙ 11　㉚ 65

㉛ 75　㉜ 112　㉝ 12　㉞ 40　㉟ 68　㊱ 99　㊲ 60　㊳ 39　㊴ 133　㊵ 96

Page 40

① 144　② 38　③ 36　④ 85　⑤ 96　⑥ 14　⑦ 40　⑧ 120　⑨ 180　⑩ 91

⑪ 126　⑫ 30　⑬ 52　⑭ 75　⑮ 40　⑯ 55　⑰ 12　⑱ 126　⑲ 128　⑳ 114

㉑ 108　㉒ 68　㉓ 77　㉔ 36　㉕ 105　㉖ 39　㉗ 80　㉘ 152　㉙ 10　㉚ 120

㉛ 13　㉜ 112　㉝ 95　㉞ 24　㉟ 90　㊱ 160　㊲ 51　㊳ 72　㊴ 44　㊵ 84

Page 41
① 20 ② 33 ③ 100 ④ 64 ⑤ 112 ⑥ 126 ⑦ 45 ⑧ 72 ⑨ 117 ⑩ 17
⑪ 54 ⑫ 96 ⑬ 70 ⑭ 26 ⑮ 153 ⑯ 15 ⑰ 140 ⑱ 11 ⑲ 76 ⑳ 60
㉑ 80 ㉒ 135 ㉓ 34 ㉔ 99 ㉕ 57 ㉖ 80 ㉗ 98 ㉘ 108 ㉙ 60 ㉚ 16
㉛ 19 ㉜ 56 ㉝ 50 ㉞ 60 ㉟ 104 ㊱ 84 ㊲ 30 ㊳ 144 ㊴ 102 ㊵ 22

Page 42
① 60 ② 26 ③ 85 ④ 19 ⑤ 64 ⑥ 144 ⑦ 84 ⑧ 30 ⑨ 99 ⑩ 84
⑪ 162 ⑫ 60 ⑬ 98 ⑭ 10 ⑮ 22 ⑯ 15 ⑰ 136 ⑱ 120 ⑲ 48 ⑳ 52
㉑ 153 ㉒ 133 ㉓ 90 ㉔ 40 ㉕ 60 ㉖ 70 ㉗ 16 ㉘ 39 ㉙ 96 ㉚ 44
㉛ 112 ㉜ 96 ㉝ 13 ㉞ 34 ㉟ 108 ㊱ 33 ㊲ 95 ㊳ 140 ㊴ 105 ㊵ 72

Page 43
① 24 ② 75 ③ 11 ④ 112 ⑤ 54 ⑥ 120 ⑦ 50 ⑧ 68 ⑨ 126 ⑩ 152
⑪ 100 ⑫ 51 ⑬ 18 ⑭ 28 ⑮ 171 ⑯ 80 ⑰ 66 ⑱ 120 ⑲ 91 ⑳ 48
㉑ 77 ㉒ 90 ㉓ 38 ㉔ 135 ㉕ 65 ㉖ 36 ㉗ 108 ㉘ 80 ㉙ 17 ㉚ 128
㉛ 104 ㉜ 126 ㉝ 12 ㉞ 55 ㉟ 42 ㊱ 102 ㊲ 20 ㊳ 144 ㊴ 76 ㊵ 30

Page 44
① 24 ② 85 ③ 10 ④ 108 ⑤ 48 ⑥ 76 ⑦ 140 ⑧ 22 ⑨ 104 ⑩ 126
⑪ 152 ⑫ 20 ⑬ 98 ⑭ 66 ⑮ 65 ⑯ 90 ⑰ 40 ⑱ 108 ⑲ 32 ⑳ 51
㉑ 80 ㉒ 126 ㉓ 135 ㉔ 60 ㉕ 99 ㉖ 14 ㉗ 96 ㉘ 34 ㉙ 80 ㉚ 39
㉛ 56 ㉜ 144 ㉝ 102 ㉞ 50 ㉟ 160 ㊱ 26 ㊲ 18 ㊳ 84 ㊴ 105 ㊵ 57

Page 45
① 100 ② 15 ③ 78 ④ 112 ⑤ 38 ⑥ 108 ⑦ 11 ⑧ 30 ⑨ 112 ⑩ 72
⑪ 12 ⑫ 20 ⑬ 114 ⑭ 70 ⑮ 144 ⑯ 44 ⑰ 117 ⑱ 60 ⑲ 119 ⑳ 60
㉑ 91 ㉒ 88 ㉓ 90 ㉔ 120 ㉕ 17 ㉖ 40 ㉗ 171 ㉘ 36 ㉙ 60 ㉚ 64
㉛ 68 ㉜ 133 ㉝ 120 ㉞ 13 ㉟ 28 ㊱ 90 ㊲ 55 ㊳ 128 ㊴ 54 ㊵ 75

Page 46
① 102 ② 40 ③ 28 ④ 140 ⑤ 108 ⑥ 15 ⑦ 88 ⑧ 96 ⑨ 54 ⑩ 65
⑪ 45 ⑫ 22 ⑬ 104 ⑭ 112 ⑮ 72 ⑯ 133 ⑰ 14 ⑱ 85 ⑲ 72 ⑳ 90
㉑ 42 ㉒ 160 ㉓ 95 ㉔ 68 ㉕ 80 ㉖ 26 ㉗ 84 ㉘ 60 ㉙ 11 ㉚ 162
㉛ 13 ㉜ 60 ㉝ 70 ㉞ 56 ㉟ 33 ㊱ 108 ㊲ 40 ㊳ 136 ㊴ 152 ㊵ 135

Page 47

① 44 ② 38 ③ 126 ④ 96 ⑤ 90 ⑥ 85 ⑦ 32 ⑧ 126 ⑨ 39 ⑩ 20
⑪ 34 ⑫ 84 ⑬ 105 ⑭ 52 ⑮ 60 ⑯ 16 ⑰ 90 ⑱ 19 ⑲ 80 ⑳ 99
㉑ 144 ㉒ 48 ㉓ 80 ㉔ 57 ㉕ 20 ㉖ 66 ㉗ 75 ㉘ 153 ㉙ 98 ㉚ 12
㉛ 10 ㉜ 120 ㉝ 77 ㉞ 36 ㉟ 78 ㊱ 70 ㊲ 64 ㊳ 36 ㊴ 180 ㊵ 76

Page 48

① 60 ② 90 ③ 52 ④ 112 ⑤ 19 ⑥ 22 ⑦ 140 ⑧ 75 ⑨ 102 ⑩ 54
⑪ 66 ⑫ 80 ⑬ 126 ⑭ 120 ⑮ 153 ⑯ 128 ⑰ 26 ⑱ 36 ⑲ 95 ⑳ 10
㉑ 78 ㉒ 98 ㉓ 48 ㉔ 108 ㉕ 45 ㉖ 72 ㉗ 152 ㉘ 50 ㉙ 40 ㉚ 17
㉛ 36 ㉜ 57 ㉝ 80 ㉞ 117 ㉟ 120 ㊱ 85 ㊲ 56 ㊳ 84 ㊴ 112 ㊵ 11

Page 49

① 180 ② 64 ③ 136 ④ 133 ⑤ 55 ⑥ 36 ⑦ 60 ⑧ 13 ⑨ 108 ⑩ 28
⑪ 48 ⑫ 65 ⑬ 88 ⑭ 162 ⑮ 84 ⑯ 30 ⑰ 51 ⑱ 32 ⑲ 70 ⑳ 20
㉑ 119 ㉒ 90 ㉓ 126 ㉔ 96 ㉕ 40 ㉖ 100 ㉗ 33 ㉘ 12 ㉙ 104 ㉚ 38
㉛ 20 ㉜ 77 ㉝ 160 ㉞ 68 ㉟ 90 ㊱ 39 ㊲ 135 ㊳ 114 ㊴ 14 ㊵ 144

Page 50

① 171 ② 10 ③ 22 ④ 84 ⑤ 90 ⑥ 54 ⑦ 112 ⑧ 144 ⑨ 68 ⑩ 100
⑪ 72 ⑫ 28 ⑬ 160 ⑭ 112 ⑮ 17 ⑯ 91 ⑰ 33 ⑱ 95 ⑲ 135 ⑳ 60
㉑ 44 ㉒ 96 ㉓ 65 ㉔ 19 ㉕ 80 ㉖ 40 ㉗ 105 ㉘ 90 ㉙ 42 ㉚ 102
㉛ 60 ㉜ 75 ㉝ 70 ㉞ 11 ㉟ 56 ㊱ 153 ㊲ 24 ㊳ 152 ㊴ 104 ㊵ 108

Page 51

① 14 ② 26 ③ 119 ④ 120 ⑤ 162 ⑥ 95 ⑦ 32 ⑧ 66 ⑨ 80 ⑩ 36
⑪ 38 ⑫ 99 ⑬ 126 ⑭ 20 ⑮ 48 ⑯ 48 ⑰ 85 ⑱ 39 ⑲ 80 ⑳ 84
㉑ 136 ㉒ 64 ㉓ 12 ㉔ 52 ㉕ 20 ㉖ 126 ㉗ 90 ㉘ 114 ㉙ 77 ㉚ 45
㉛ 30 ㉜ 144 ㉝ 98 ㉞ 34 ㉟ 180 ㊱ 55 ㊲ 16 ㊳ 60 ㊴ 72 ㊵ 13

Page 52

① 152 ② 85 ③ 84 ④ 99 ⑤ 16 ⑥ 126 ⑦ 39 ⑧ 120 ⑨ 40 ⑩ 40
⑪ 72 ⑫ 78 ⑬ 60 ⑭ 135 ⑮ 50 ⑯ 108 ⑰ 98 ⑱ 38 ⑲ 128 ⑳ 17
㉑ 56 ㉒ 33 ㉓ 24 ㉔ 95 ㉕ 30 ㉖ 120 ㉗ 144 ㉘ 136 ㉙ 91 ㉚ 10
㉛ 140 ㉜ 32 ㉝ 153 ㉞ 70 ㉟ 52 ㊱ 80 ㊲ 66 ㊳ 57 ㊴ 36 ㊵ 18

Page 53
① 65 ② 72 ③ 90 ④ 48 ⑤ 144 ⑥ 38 ⑦ 90 ⑧ 14 ⑨ 80 ⑩ 77
⑪ 114 ⑫ 112 ⑬ 162 ⑭ 100 ⑮ 44 ⑯ 105 ⑰ 20 ⑱ 84 ⑲ 51 ⑳ 13
㉑ 30 ㉒ 60 ㉓ 55 ㉔ 48 ㉕ 102 ㉖ 104 ㉗ 36 ㉘ 19 ㉙ 126 ㉚ 112
㉛ 119 ㉜ 54 ㉝ 117 ㉞ 60 ㉟ 160 ㊱ 28 ㊲ 75 ㊳ 64 ㊴ 11 ㊵ 60

Page 54
① 40 ② 256 ③ 112 ④ 228 ⑤ 36 ⑥ 187 ⑦ 117 ⑧ 285 ⑨ 100 ⑩ 324
⑪ 114 ⑫ 156 ⑬ 110 ⑭ 165 ⑮ 13 ⑯ 210 ⑰ 140 ⑱ 252 ⑲ 32 ⑳ 340
㉑ 162 ㉒ 240 ㉓ 65 ㉔ 190 ㉕ 64 ㉖ 342 ㉗ 96 ㉘ 289 ㉙ 45 ㉚ 176
㉛ 105 ㉜ 143 ㉝ 96 ㉞ 260 ㉟ 200 ㊱ 198 ㊲ 34 ㊳ 196 ㊴ 10 ㊵ 180

Page 55
① 104 ② 180 ③ 72 ④ 192 ⑤ 126 ⑥ 192 ⑦ 33 ⑧ 323 ⑨ 85 ⑩ 380
⑪ 40 ⑫ 154 ⑬ 19 ⑭ 221 ⑮ 66 ⑯ 210 ⑰ 100 ⑱ 320 ⑲ 126 ⑳ 195
㉑ 48 ㉒ 306 ㉓ 51 ㉔ 266 ㉕ 50 ㉖ 320 ㉗ 144 ㉘ 156 ㉙ 88 ㉚ 270
㉛ 119 ㉜ 247 ㉝ 120 ㉞ 240 ㉟ 15 ㊱ 182 ㊲ 36 ㊳ 121 ㊴ 84 ㊵ 200

Page 56
① 98 ② 154 ③ 100 ④ 289 ⑤ 130 ⑥ 270 ⑦ 80 ⑧ 132 ⑨ 162 ⑩ 285
⑪ 34 ⑫ 156 ⑬ 15 ⑭ 192 ⑮ 40 ⑯ 260 ⑰ 108 ⑱ 361 ⑲ 33 ⑳ 320
㉑ 152 ㉒ 306 ㉓ 90 ㉔ 154 ㉕ 77 ㉖ 255 ㉗ 65 ㉘ 288 ㉙ 120 ㉚ 210
㉛ 72 ㉜ 180 ㉝ 22 ㉞ 200 ㉟ 18 ㊱ 304 ㊲ 48 ㊳ 380 ㊴ 56 ㊵ 169

Page 57
① 50 ② 210 ③ 133 ④ 182 ⑤ 160 ⑥ 187 ⑦ 150 ⑧ 306 ⑨ 144 ⑩ 198
⑪ 54 ⑫ 320 ⑬ 68 ⑭ 192 ⑮ 30 ⑯ 228 ⑰ 14 ⑱ 220 ⑲ 114 ⑳ 130
㉑ 91 ㉒ 342 ㉓ 160 ㉔ 220 ㉕ 126 ㉖ 252 ㉗ 88 ㉘ 170 ㉙ 75 ㉚ 180
㉛ 96 ㉜ 204 ㉝ 13 ㉞ 143 ㉟ 48 ㊱ 190 ㊲ 57 ㊳ 240 ㊴ 40 ㊵ 280

Page 58
① 190 ② 270 ③ 160 ④ 304 ⑤ 77 ⑥ 130 ⑦ 60 ⑧ 156 ⑨ 42 ⑩ 270
⑪ 144 ⑫ 154 ⑬ 60 ⑭ 221 ⑮ 24 ⑯ 320 ⑰ 56 ⑱ 300 ⑲ 18 ⑳ 187
㉑ 75 ㉒ 266 ㉓ 36 ㉔ 228 ㉕ 180 ㉖ 288 ㉗ 88 ㉘ 221 ㉙ 91 ㉚ 150
㉛ 52 ㉜ 140 ㉝ 162 ㉞ 132 ㉟ 84 ㊱ 255 ㊲ 17 ㊳ 400 ㊴ 38 ㊵ 176

Page 59
① 96　② 342　③ 150　④ 165　⑤ 100　⑥ 342　⑦ 70　⑧ 208　⑨ 51　⑩ 168
⑪ 14　⑫ 340　⑬ 32　⑭ 238　⑮ 90　⑯ 260　⑰ 114　⑱ 198　⑲ 60　⑳ 192
㉑ 110　㉒ 195　㉓ 119　㉔ 240　㉕ 57　㉖ 210　㉗ 90　㉘ 228　㉙ 80　㉚ 234
㉛ 68　㉜ 224　㉝ 66　㉞ 288　㉟ 26　㊱ 240　㊲ 15　㊳ 170　㊴ 180　㊵ 209

Page 60
① 90　② 360　③ 42　④ 342　⑤ 130　⑥ 266　⑦ 15　⑧ 204　⑨ 55　⑩ 240
⑪ 72　⑫ 260　⑬ 133　⑭ 272　⑮ 90　⑯ 182　⑰ 22　⑱ 272　⑲ 160　⑳ 132
㉑ 16　㉒ 209　㉓ 75　㉔ 120　㉕ 180　㉖ 270　㉗ 39　㉘ 168　㉙ 170　㉚ 342
㉛ 34　㉜ 380　㉝ 80　㉞ 165　㉟ 77　㊱ 256　㊲ 96　㊳ 238　㊴ 60　㊵ 169

Page 61
① 45　② 150　③ 144　④ 234　⑤ 14　⑥ 380　⑦ 190　⑧ 252　⑨ 60　⑩ 132
⑪ 88　⑫ 224　⑬ 108　⑭ 240　⑮ 76　⑯ 289　⑰ 70　⑱ 220　⑲ 32　⑳ 195
㉑ 117　㉒ 224　㉓ 120　㉔ 168　㉕ 50　㉖ 198　㉗ 20　㉘ 285　㉙ 57　㉚ 255
㉛ 24　㉜ 360　㉝ 91　㉞ 260　㉟ 102　㊱ 255　㊲ 128　㊳ 304　㊴ 56　㊵ 110

Page 62
① 90　② 306　③ 70　④ 192　⑤ 56　⑥ 270　⑦ 99　⑧ 209　⑨ 60　⑩ 272
⑪ 12　⑫ 168　⑬ 36　⑭ 266　⑮ 110　⑯ 190　⑰ 100　⑱ 208　⑲ 136　⑳ 260
㉑ 144　㉒ 320　㉓ 33　㉔ 165　㉕ 102　㉖ 204　㉗ 98　㉘ 195　㉙ 76　㉚ 324
㉛ 95　㉜ 216　㉝ 17　㉞ 220　㉟ 40　㊱ 224　㊲ 104　㊳ 130　㊴ 150　㊵ 266

Page 63
① 77　② 255　③ 96　④ 252　⑤ 90　⑥ 272　⑦ 72　⑧ 180　⑨ 39　⑩ 220
⑪ 160　⑫ 140　⑬ 120　⑭ 156　⑮ 18　⑯ 247　⑰ 30　⑱ 340　⑲ 80　⑳ 209
㉑ 84　㉒ 240　㉓ 52　㉔ 110　㉕ 45　㉖ 360　㉗ 153　㉘ 176　㉙ 126　㉚ 323
㉛ 65　㉜ 144　㉝ 28　㉞ 323　㉟ 190　㊱ 143　㊲ 128　㊳ 252　㊴ 10　㊵ 300

Page 64
① 128　② 280　③ 50　④ 176　⑤ 38　⑥ 238　⑦ 180　⑧ 204　⑨ 60　⑩ 247
⑪ 11　⑫ 209　⑬ 119　⑭ 144　⑮ 72　⑯ 180　⑰ 120　⑱ 195　⑲ 126　⑳ 195
㉑ 130　㉒ 320　㉓ 54　㉔ 272　㉕ 112　㉖ 209　㉗ 95　㉘ 210　㉙ 24　㉚ 340
㉛ 72　㉜ 187　㉝ 14　㉞ 234　㉟ 140　㊱ 156　㊲ 135　㊳ 150　㊴ 64　㊵ 342

Page 65

① 90 ② 304 ③ 104 ④ 380 ⑤ 100 ⑥ 224 ⑦ 34 ⑧ 154 ⑨ 45 ⑩ 340
⑪ 180 ⑫ 120 ⑬ 44 ⑭ 165 ⑮ 17 ⑯ 180 ⑰ 112 ⑱ 182 ⑲ 78 ⑳ 324
㉑ 152 ㉒ 182 ㉓ 30 ㉔ 170 ㉕ 48 ㉖ 400 ㉗ 140 ㉘ 288 ㉙ 85 ㉚ 228
㉛ 90 ㉜ 121 ㉝ 133 ㉞ 252 ㉟ 48 ㊱ 270 ㊲ 117 ㊳ 204 ㊴ 10 ㊵ 208

Page 66

① 136 ② 216 ③ 32 ④ 154 ⑤ 78 ⑥ 260 ⑦ 70 ⑧ 216 ⑨ 150 ⑩ 200
⑪ 44 ⑫ 221 ⑬ 180 ⑭ 342 ⑮ 14 ⑯ 240 ⑰ 105 ⑱ 160 ⑲ 36 ⑳ 209
㉑ 50 ㉒ 210 ㉓ 140 ㉔ 204 ㉕ 96 ㉖ 220 ㉗ 26 ㉘ 247 ㉙ 108 ㉚ 360
㉛ 126 ㉜ 340 ㉝ 48 ㉞ 154 ㉟ 135 ㊱ 190 ㊲ 57 ㊳ 256 ㊴ 17 ㊵ 195

Page 67

① 28 ② 340 ③ 80 ④ 234 ⑤ 80 ⑥ 168 ⑦ 120 ⑧ 143 ⑨ 190 ⑩ 180
⑪ 45 ⑫ 304 ⑬ 11 ⑭ 323 ⑮ 80 ⑯ 288 ⑰ 153 ⑱ 132 ⑲ 70 ⑳ 210
㉑ 104 ㉒ 130 ㉓ 114 ㉔ 192 ㉕ 170 ㉖ 270 ㉗ 60 ㉘ 196 ㉙ 40 ㉚ 360
㉛ 133 ㉜ 187 ㉝ 117 ㉞ 180 ㉟ 18 ㊱ 224 ㊲ 30 ㊳ 380 ㊴ 64 ㊵ 187

Page 68

① 12 ② 150 ③ 133 ④ 196 ⑤ 22 ⑥ 320 ⑦ 162 ⑧ 169 ⑨ 64 ⑩ 306
⑪ 112 ⑫ 121 ⑬ 200 ⑭ 156 ⑮ 54 ⑯ 380 ⑰ 80 ⑱ 323 ⑲ 60 ⑳ 255
㉑ 153 ㉒ 224 ㉓ 72 ㉔ 156 ㉕ 10 ㉖ 252 ㉗ 77 ㉘ 240 ㉙ 26 ㉚ 300
㉛ 65 ㉜ 220 ㉝ 80 ㉞ 306 ㉟ 160 ㊱ 204 ㊲ 90 ㊳ 361 ㊴ 36 ㊵ 220

Page 69

① 126 ② 216 ③ 17 ④ 165 ⑤ 171 ⑥ 140 ⑦ 40 ⑧ 224 ⑨ 60 ⑩ 208
⑪ 96 ⑫ 228 ⑬ 42 ⑭ 165 ⑮ 160 ⑯ 247 ⑰ 120 ⑱ 170 ⑲ 85 ⑳ 360
㉑ 11 ㉒ 272 ㉓ 30 ㉔ 247 ㉕ 48 ㉖ 240 ㉗ 90 ㉘ 252 ㉙ 140 ㉚ 234
㉛ 75 ㉜ 154 ㉝ 110 ㉞ 200 ㉟ 39 ㊱ 342 ㊲ 102 ㊳ 240 ㊴ 152 ㊵ 204

Page 70

① 160 ② 224 ③ 90 ④ 285 ⑤ 30 ⑥ 340 ⑦ 22 ⑧ 168 ⑨ 108 ⑩ 165
⑪ 19 ⑫ 180 ⑬ 170 ⑭ 266 ⑮ 77 ⑯ 288 ⑰ 72 ⑱ 255 ⑲ 64 ⑳ 169
㉑ 30 ㉒ 180 ㉓ 99 ㉔ 240 ㉕ 128 ㉖ 209 ㉗ 50 ㉘ 260 ㉙ 42 ㉚ 238
㉛ 84 ㉜ 306 ㉝ 16 ㉞ 143 ㉟ 120 ㊱ 285 ㊲ 52 ㊳ 306 ㊴ 140 ㊵ 160

Page 71

① 55 ② 220 ③ 120 ④ 140 ⑤ 36 ⑥ 240 ⑦ 51 ⑧ 380 ⑨ 117 ⑩ 144
⑪ 48 ⑫ 342 ⑬ 133 ⑭ 234 ⑮ 17 ⑯ 238 ⑰ 200 ⑱ 208 ⑲ 90 ⑳ 176
㉑ 80 ㉒ 300 ㉓ 39 ㉔ 216 ㉕ 180 ㉖ 168 ㉗ 32 ㉘ 165 ㉙ 85 ㉚ 154
㉛ 78 ㉜ 342 ㉝ 100 ㉞ 256 ㉟ 98 ㊱ 187 ㊲ 60 ㊳ 323 ㊴ 18 ㊵ 260

Page 72

① 64 ② 216 ③ 200 ④ 143 ⑤ 30 ⑥ 238 ⑦ 84 ⑧ 143 ⑨ 135 ⑩ 240
⑪ 22 ⑫ 170 ⑬ 17 ⑭ 195 ⑮ 112 ⑯ 380 ⑰ 105 ⑱ 192 ⑲ 90 ⑳ 342
㉑ 72 ㉒ 195 ㉓ 126 ㉔ 176 ㉕ 72 ㉖ 220 ㉗ 100 ㉘ 266 ㉙ 39 ㉚ 204
㉛ 91 ㉜ 289 ㉝ 36 ㉞ 252 ㉟ 15 ㊱ 180 ㊲ 95 ㊳ 320 ㊴ 128 ㊵ 190

Page 73

① 140 ② 320 ③ 48 ④ 120 ⑤ 120 ⑥ 234 ⑦ 51 ⑧ 154 ⑨ 171 ⑩ 165
⑪ 75 ⑫ 300 ⑬ 88 ⑭ 323 ⑮ 34 ⑯ 208 ⑰ 16 ⑱ 324 ⑲ 84 ⑳ 266
㉑ 40 ㉒ 168 ㉓ 57 ㉔ 220 ㉕ 144 ㉖ 180 ㉗ 108 ㉘ 182 ㉙ 170 ㉚ 260
㉛ 133 ㉜ 187 ㉝ 10 ㉞ 342 ㉟ 104 ㊱ 224 ㊲ 60 ㊳ 255 ㊴ 40 ㊵ 288

Page 74

① 84 ② 209 ③ 160 ④ 130 ⑤ 60 ⑥ 252 ⑦ 18 ⑧ 323 ⑨ 96 ⑩ 221
⑪ 100 ⑫ 225 ⑬ 126 ⑭ 272 ⑮ 54 ⑯ 240 ⑰ 80 ⑱ 182 ⑲ 38 ⑳ 220
㉑ 13 ㉒ 208 ㉓ 108 ㉔ 228 ㉕ 133 ㉖ 170 ㉗ 120 ㉘ 198 ㉙ 68 ㉚ 154
㉛ 85 ㉜ 210 ㉝ 190 ㉞ 360 ㉟ 144 ㊱ 208 ㊲ 22 ㊳ 280 ㊴ 36 ㊵ 180

Page 75

① 144 ② 204 ③ 91 ④ 165 ⑤ 20 ⑥ 247 ⑦ 56 ⑧ 180 ⑨ 66 ⑩ 304
⑪ 32 ⑫ 320 ⑬ 30 ⑭ 165 ⑮ 140 ⑯ 238 ⑰ 108 ⑱ 380 ⑲ 65 ⑳ 216
㉑ 105 ㉒ 234 ㉓ 44 ㉔ 380 ㉕ 72 ㉖ 176 ㉗ 19 ㉘ 234 ㉙ 112 ㉚ 289
㉛ 55 ㉜ 150 ㉝ 135 ㉞ 228 ㉟ 51 ㊱ 252 ㊲ 26 ㊳ 224 ㊴ 200 ㊵ 240

Page 76

① 126 ② 300 ③ 77 ④ 240 ⑤ 45 ⑥ 323 ⑦ 108 ⑧ 130 ⑨ 13 ⑩ 304
⑪ 48 ⑫ 180 ⑬ 34 ⑭ 110 ⑮ 144 ⑯ 187 ⑰ 65 ⑱ 266 ⑲ 200 ⑳ 288
㉑ 114 ㉒ 260 ㉓ 18 ㉔ 182 ㉕ 180 ㉖ 192 ㉗ 105 ㉘ 304 ㉙ 30 ㉚ 255
㉛ 50 ㉜ 204 ㉝ 80 ㉞ 324 ㉟ 26 ㊱ 209 ㊲ 160 ㊳ 154 ㊴ 112 ㊵ 255

Page 77
① 126 ② 224 ③ 171 ④ 225 ⑤ 66 ⑥ 400 ⑦ 51 ⑧ 228 ⑨ 16 ⑩ 169
⑪ 130 ⑫ 121 ⑬ 96 ⑭ 192 ⑮ 68 ⑯ 140 ⑰ 28 ⑱ 360 ⑲ 95 ⑳ 306
㉑ 135 ㉒ 361 ㉓ 48 ㉔ 143 ㉕ 14 ㉖ 195 ㉗ 120 ㉘ 360 ㉙ 119 ㉚ 160
㉛ 80 ㉜ 144 ㉝ 30 ㉞ 340 ㉟ 80 ㊱ 252 ㊲ 190 ㊳ 187 ㊴ 44 ㊵ 252

Page 78
① 90 ② 208 ③ 120 ④ 196 ⑤ 144 ⑥ 380 ⑦ 40 ⑧ 143 ⑨ 38 ⑩ 306
⑪ 98 ⑫ 360 ⑬ 20 ⑭ 156 ⑮ 90 ⑯ 180 ⑰ 57 ⑱ 272 ⑲ 80 ⑳ 187
㉑ 68 ㉒ 266 ㉓ 20 ㉔ 165 ㉕ 96 ㉖ 252 ㉗ 180 ㉘ 209 ㉙ 104 ㉚ 260
㉛ 39 ㉜ 400 ㉝ 112 ㉞ 170 ㉟ 19 ㊱ 204 ㊲ 55 ㊳ 192 ㊴ 135 ㊵ 270

Page 79
① 100 ② 270 ③ 102 ④ 234 ⑤ 48 ⑥ 224 ⑦ 160 ⑧ 266 ⑨ 22 ⑩ 209
⑪ 95 ⑫ 144 ⑬ 126 ⑭ 220 ⑮ 140 ⑯ 208 ⑰ 15 ⑱ 272 ⑲ 51 ⑳ 150
㉑ 108 ㉒ 323 ㉓ 88 ㉔ 132 ㉕ 30 ㉖ 247 ㉗ 64 ㉘ 140 ㉙ 200 ㉚ 234
㉛ 33 ㉜ 280 ㉝ 18 ㉞ 240 ㉟ 117 ㊱ 160 ㊲ 85 ㊳ 240 ㊴ 84 ㊵ 255

Page 80
① 150 ② 234 ③ 304 ④ 150 ⑤ 342 ⑥ 121 ⑦ 238 ⑧ 280 ⑨ 240 ⑩ 192
⑪ 280 ⑫ 187 ⑬ 300 ⑭ 180 ⑮ 361 ⑯ 180 ⑰ 192 ⑱ 169 ⑲ 180 ⑳ 187
㉑ 228 ㉒ 143 ㉓ 360 ㉔ 132 ㉕ 320 ㉖ 221 ㉗ 160 ㉘ 190 ㉙ 306 ㉚ 196
㉛ 228 ㉜ 200 ㉝ 156 ㉞ 240 ㉟ 198 ㊱ 140 ㊲ 289 ㊳ 209 ㊴ 300 ㊵ 234

Page 81
① 195 ② 252 ③ 100 ④ 224 ⑤ 204 ⑥ 304 ⑦ 220 ⑧ 216 ⑨ 220 ⑩ 285
⑪ 132 ⑫ 272 ⑬ 165 ⑭ 266 ⑮ 130 ⑯ 182 ⑰ 380 ⑱ 288 ⑲ 150 ⑳ 340
㉑ 360 ㉒ 380 ㉓ 272 ㉔ 306 ㉕ 120 ㉖ 225 ㉗ 182 ㉘ 208 ㉙ 154 ㉚ 120
㉛ 143 ㉜ 140 ㉝ 156 ㉞ 165 ㉟ 256 ㊱ 323 ㊲ 216 ㊳ 210 ㊴ 342 ㊵ 340

Page 82
① 234 ② 320 ③ 168 ④ 165 ⑤ 216 ⑥ 143 ⑦ 190 ⑧ 320 ⑨ 190 ⑩ 210
⑪ 100 ⑫ 187 ⑬ 260 ⑭ 360 ⑮ 252 ⑯ 154 ⑰ 228 ⑱ 272 ⑲ 180 ⑳ 247
㉑ 196 ㉒ 221 ㉓ 120 ㉔ 209 ㉕ 240 ㉖ 304 ㉗ 225 ㉘ 324 ㉙ 340 ㉚ 160
㉛ 342 ㉜ 300 ㉝ 224 ㉞ 195 ㉟ 260 ㊱ 288 ㊲ 323 ㊳ 154 ㊴ 110 ㊵ 204

Page 83

① 187 ② 192 ③ 270 ④ 120 ⑤ 238 ⑥ 270 ⑦ 130 ⑧ 380 ⑨ 220 ⑩ 182
⑪ 182 ⑫ 285 ⑬ 121 ⑭ 224 ⑮ 240 ⑯ 170 ⑰ 360 ⑱ 144 ⑲ 234 ⑳ 170
㉑ 400 ㉒ 140 ㉓ 204 ㉔ 228 ㉕ 252 ㉖ 143 ㉗ 256 ㉘ 255 ㉙ 110 ㉚ 285
㉛ 176 ㉜ 150 ㉝ 323 ㉞ 169 ㉟ 180 ㊱ 306 ㊲ 168 ㊳ 176 ㊴ 280 ㊵ 380

Page 84

① 130 ② 204 ③ 195 ④ 196 ⑤ 342 ⑥ 300 ⑦ 176 ⑧ 204 ⑨ 342 ⑩ 110
⑪ 209 ⑫ 240 ⑬ 170 ⑭ 221 ⑮ 270 ⑯ 154 ⑰ 234 ⑱ 240 ⑲ 266 ⑳ 320
㉑ 165 ㉒ 400 ㉓ 361 ㉔ 270 ㉕ 221 ㉖ 192 ㉗ 140 ㉘ 234 ㉙ 272 ㉚ 132
㉛ 324 ㉜ 238 ㉝ 132 ㉞ 100 ㉟ 340 ㊱ 156 ㊲ 256 ㊳ 225 ㊴ 266 ㊵ 380

Page 85

① 280 ② 228 ③ 182 ④ 143 ⑤ 176 ⑥ 180 ⑦ 380 ⑧ 306 ⑨ 255 ⑩ 150
⑪ 220 ⑫ 160 ⑬ 210 ⑭ 180 ⑮ 168 ⑯ 220 ⑰ 306 ⑱ 247 ⑲ 130 ⑳ 304
㉑ 289 ㉒ 285 ㉓ 208 ㉔ 304 ㉕ 143 ㉖ 140 ㉗ 144 ㉘ 200 ㉙ 165 ㉚ 252
㉛ 180 ㉜ 154 ㉝ 360 ㉞ 200 ㉟ 130 ㊱ 288 ㊲ 209 ㊳ 168 ㊴ 255 ㊵ 272

Page 86

① 180 ② 288 ③ 170 ④ 238 ⑤ 120 ⑥ 380 ⑦ 252 ⑧ 121 ⑨ 260 ⑩ 247
⑪ 360 ⑫ 285 ⑬ 165 ⑭ 216 ⑮ 204 ⑯ 182 ⑰ 130 ⑱ 256 ⑲ 170 ⑳ 280
㉑ 221 ㉒ 320 ㉓ 200 ㉔ 247 ㉕ 110 ㉖ 224 ㉗ 323 ㉘ 144 ㉙ 270 ㉚ 198
㉛ 156 ㉜ 187 ㉝ 208 ㉞ 285 ㉟ 360 ㊱ 132 ㊲ 210 ㊳ 323 ㊴ 280 ㊵ 160

Page 87

① 224 ② 110 ③ 204 ④ 180 ⑤ 195 ⑥ 228 ⑦ 400 ⑧ 234 ⑨ 209 ⑩ 255
⑪ 260 ⑫ 266 ⑬ 266 ⑭ 187 ⑮ 272 ⑯ 288 ⑰ 132 ⑱ 100 ⑲ 180 ⑳ 300
㉑ 198 ㉒ 340 ㉓ 150 ㉔ 140 ㉕ 156 ㉖ 210 ㉗ 176 ㉘ 304 ㉙ 342 ㉚ 221
㉛ 304 ㉜ 306 ㉝ 208 ㉞ 300 ㉟ 190 ㊱ 180 ㊲ 130 ㊳ 154 ㊴ 306 ㊵ 154

Page 88

① 176 ② 220 ③ 342 ④ 238 ⑤ 288 ⑥ 100 ⑦ 285 ⑧ 255 ⑨ 168 ⑩ 260
⑪ 228 ⑫ 120 ⑬ 240 ⑭ 121 ⑮ 324 ⑯ 340 ⑰ 266 ⑱ 260 ⑲ 224 ⑳ 150
㉑ 360 ㉒ 130 ㉓ 192 ㉔ 140 ㉕ 285 ㉖ 300 ㉗ 182 ㉘ 198 ㉙ 132 ㉚ 323
㉛ 252 ㉜ 210 ㉝ 400 ㉞ 208 ㉟ 110 ㊱ 187 ㊲ 180 ㊳ 180 ㊴ 204 ㊵ 208

Page 89

①221 ②272 ③154 ④361 ⑤240 ⑥234 ⑦120 ⑧154 ⑨150 ⑩288
⑪200 ⑫195 ⑬170 ⑭306 ⑮280 ⑯247 ⑰270 ⑱304 ⑲176 ⑳144
㉑165 ㉒216 ㉓228 ㉔240 ㉕220 ㉖272 ㉗340 ㉘169 ㉙190 ㉚196
㉛200 ㉜266 ㉝182 ㉞160 ㉟247 ㊱216 ㊲320 ㊳289 ㊴198 ㊵225

Page 90

①228 ②280 ③260 ④216 ⑤240 ⑥180 ⑦238 ⑧323 ⑨143 ⑩240
⑪182 ⑫100 ⑬204 ⑭266 ⑮195 ⑯192 ⑰220 ⑱220 ⑲288 ⑳306
㉑208 ㉒198 ㉓208 ㉔110 ㉕340 ㉖340 ㉗270 ㉘285 ㉙140 ㉚266
㉛165 ㉜306 ㉝320 ㉞180 ㉟252 ㊱361 ㊲170 ㊳130 ㊴156 ㊵176

Page 91

①132 ②304 ③342 ④280 ⑤160 ⑥225 ⑦234 ⑧154 ⑨170 ⑩156
⑪288 ⑫255 ⑬120 ⑭247 ⑮360 ⑯154 ⑰255 ⑱320 ⑲228 ⑳130
㉑238 ㉒169 ㉓200 ㉔272 ㉕304 ㉖144 ㉗380 ㉘165 ㉙140 ㉚198
㉛200 ㉜252 ㉝121 ㉞216 ㉟300 ㊱150 ㊲256 ㊳228 ㊴182 ㊵289

Page 92

①270 ②100 ③224 ④216 ⑤380 ⑥180 ⑦380 ⑧121 ⑨221 ⑩224
⑪323 ⑫255 ⑬198 ⑭150 ⑮304 ⑯288 ⑰182 ⑱130 ⑲240 ⑳240
㉑256 ㉒156 ㉓340 ㉔195 ㉕154 ㉖200 ㉗168 ㉘285 ㉙170 ㉚209
㉛247 ㉜132 ㉝160 ㉞252 ㉟120 ㊱165 ㊲340 ㊳240 ㊴306 ㊵260

Page 93

①234 ②220 ③180 ④266 ⑤272 ⑥266 ⑦204 ⑧130 ⑨165 ⑩288
⑪192 ⑫280 ⑬270 ⑭176 ⑮120 ⑯247 ⑰209 ⑱280 ⑲270 ⑳289
㉑110 ㉒272 ㉓238 ㉔400 ㉕240 ㉖324 ㉗110 ㉘182 ㉙285 ㉚156
㉛150 ㉜228 ㉝169 ㉞216 ㉟196 ㊱323 ㊲320 ㊳198 ㊴160 ㊵220

Page 94

①360 ②190 ③195 ④340 ⑤192 ⑥247 ⑦180 ⑧204 ⑨154 ⑩240
⑪198 ⑫143 ⑬216 ⑭380 ⑮156 ⑯272 ⑰210 ⑱140 ⑲320 ⑳190
㉑208 ㉒266 ㉓176 ㉔182 ㉕180 ㉖100 ㉗300 ㉘240 ㉙209 ㉚306
㉛168 ㉜240 ㉝160 ㉞270 ㉟361 ㊱340 ㊲110 ㊳169 ㊴187 ㊵224

Page 95

①238 ②272 ③400 ④270 ⑤176 ⑥180 ⑦209 ⑧266 ⑨156 ⑩234
⑪304 ⑫150 ⑬221 ⑭221 ⑮200 ⑯252 ⑰144 ⑱240 ⑲360 ⑳121
㉑228 ㉒143 ㉓165 ㉔160 ㉕320 ㉖306 ㉗170 ㉘210 ㉙234 ㉚280
㉛130 ㉜360 ㉝196 ㉞342 ㉟225 ㊱132 ㊲256 ㊳289 ㊴247 ㊵120

Page 96

①228 ②180 ③323 ④182 ⑤154 ⑥220 ⑦200 ⑧180 ⑨234 ⑩256
⑪180 ⑫165 ⑬144 ⑭190 ⑮238 ⑯224 ⑰247 ⑱306 ⑲143 ⑳400
㉑272 ㉒340 ㉓198 ㉔143 ㉕228 ㉖360 ㉗208 ㉘266 ㉙150 ㉚150
㉛182 ㉜156 ㉝288 ㉞192 ㉟200 ㊱285 ㊲300 ㊳187 ㊴252 ㊵187

Page 97

①238 ②165 ③247 ④190 ⑤240 ⑥224 ⑦360 ⑧130 ⑨132 ⑩204
⑪320 ⑫320 ⑬154 ⑭255 ⑮234 ⑯170 ⑰168 ⑱209 ⑲228 ⑳270
㉑120 ㉒306 ㉓285 ㉔220 ㉕304 ㉖168 ㉗270 ㉘272 ㉙110 ㉚169
㉛198 ㉜130 ㉝221 ㉞240 ㉟304 ㊱210 ㊲176 ㊳210 ㊴170 ㊵240

Page 98

①196 ②288 ③240 ④285 ⑤306 ⑥130 ⑦132 ⑧143 ⑨200 ⑩285
⑪110 ⑫208 ⑬182 ⑭224 ⑮216 ⑯225 ⑰400 ⑱306 ⑲323 ⑳120
㉑180 ㉒170 ㉓170 ㉔260 ㉕260 ㉖216 ㉗361 ㉘252 ㉙256 ㉚121
㉛360 ㉜247 ㉝270 ㉞266 ㉟160 ㊱154 ㊲192 ㊳156 ㊴150 ㊵289

Page 99

①255 ②187 ③266 ④220 ⑤240 ⑥342 ⑦100 ⑧320 ⑨169 ⑩168
⑪150 ⑫228 ⑬195 ⑭132 ⑮342 ⑯187 ⑰234 ⑱340 ⑲196 ⑳160
㉑208 ㉒120 ㉓320 ㉔204 ㉕210 ㉖210 ㉗198 ㉘323 ㉙143 ㉚380
㉛234 ㉜209 ㉝340 ㉞140 ㉟380 ㊱288 ㊲255 ㊳165 ㊴192 ㊵156

Page 100

①12 ②35 ③6 ④30 ⑤24 ⑥32 ⑦81 ⑧56 ⑨6 ⑩5
⑪12 ⑫3 ⑬72 ⑭63 ⑮24 ⑯14 ⑰45 ⑱8 ⑲21 ⑳48
㉑36 ㉒5 ㉓18 ㉔36 ㉕63 ㉖32 ㉗8 ㉘16 ㉙25 ㉚6
㉛40 ㉜16 ㉝49 ㉞2 ㉟10 ㊱36 ㊲3 ㊳45 ㊴7 ㊵12

Page 101

①4 ②21 ③54 ④54 ⑤20 ⑥1 ⑦7 ⑧15 ⑨64 ⑩8

⑪42 ⑫10 ⑬40 ⑭2 ⑮8 ⑯72 ⑰24 ⑱30 ⑲27 ⑳27

㉑4 ㉒16 ㉓28 ㉔9 ㉕9 ㉖14 ㉗42 ㉘6 ㉙18 ㉚20

㉛35 ㉜18 ㉝9 ㉞56 ㉟12 ㊱15 ㊲24 ㊳28 ㊴4 ㊵18

Page 102

①12 ②36 ③36 ④35 ⑤4 ⑥7 ⑦9 ⑧8 ⑨48 ⑩45

⑪16 ⑫20 ⑬24 ⑭3 ⑮42 ⑯9 ⑰21 ⑱8 ⑲16 ⑳40

㉑6 ㉒42 ㉓30 ㉔3 ㉕12 ㉖16 ㉗35 ㉘63 ㉙63 ㉚12

㉛72 ㉜2 ㉝4 ㉞54 ㉟56 ㊱10 ㊲36 ㊳27 ㊴5 ㊵4

Page 103

①72 ②6 ③15 ④6 ⑤9 ⑥30 ⑦24 ⑧28 ⑨56 ⑩18

⑪20 ⑫81 ⑬49 ⑭40 ⑮48 ⑯21 ⑰5 ⑱18 ⑲12 ⑳18

㉑10 ㉒64 ㉓1 ㉔24 ㉕18 ㉖32 ㉗2 ㉘25 ㉙27 ㉚14

㉛7 ㉜8 ㉝15 ㉞28 ㉟45 ㊱54 ㊲32 ㊳8 ㊴6 ㊵24

Page 104

①4 ②72 ③36 ④7 ⑤20 ⑥15 ⑦6 ⑧16 ⑨6 ⑩48

⑪30 ⑫54 ⑬24 ⑭4 ⑮5 ⑯8 ⑰21 ⑱18 ⑲81 ⑳8

㉑8 ㉒49 ㉓1 ㉔9 ㉕27 ㉖32 ㉗30 ㉘20 ㉙56 ㉚35

㉛64 ㉜12 ㉝18 ㉞28 ㉟42 ㊱5 ㊲7 ㊳16 ㊴12 ㊵27

Page 105

①24 ②2 ③2 ④15 ⑤24 ⑥42 ⑦63 ⑧63 ⑨40 ⑩16

⑪9 ⑫32 ⑬35 ⑭36 ⑮56 ⑯10 ⑰3 ⑱40 ⑲18 ⑳3

㉑24 ㉒48 ㉓6 ㉔9 ㉕14 ㉖54 ㉗10 ㉘6 ㉙12 ㉚28

㉛14 ㉜45 ㉝18 ㉞45 ㉟4 ㊱8 ㊲72 ㊳36 ㊴21 ㊵12

Page 106

①32 ②27 ③14 ④56 ⑤4 ⑥20 ⑦30 ⑧3 ⑨72 ⑩18

⑪9 ⑫42 ⑬18 ⑭5 ⑮40 ⑯2 ⑰42 ⑱36 ⑲63 ⑳24

㉑20 ㉒14 ㉓64 ㉔24 ㉕54 ㉖12 ㉗30 ㉘10 ㉙21 ㉚2

㉛9 ㉜16 ㉝9 ㉞4 ㉟63 ㊱8 ㊲16 ㊳15 ㊴6 ㊵28

Page 107

① 18 ② 8 ③ 40 ④ 6 ⑤ 12 ⑥ 12 ⑦ 18 ⑧ 49 ⑨ 15 ⑩ 8
⑪ 72 ⑫ 6 ⑬ 35 ⑭ 54 ⑮ 6 ⑯ 25 ⑰ 32 ⑱ 3 ⑲ 35 ⑳ 48
㉑ 24 ㉒ 27 ㉓ 4 ㉔ 56 ㉕ 10 ㉖ 81 ㉗ 1 ㉘ 48 ㉙ 12 ㉚ 28
㉛ 7 ㉜ 7 ㉝ 36 ㉞ 45 ㉟ 16 ㊱ 24 ㊲ 21 ㊳ 36 ㊴ 8 ㊵ 45

Page 108

① 2 ② 14 ③ 48 ④ 20 ⑤ 3 ⑥ 63 ⑦ 20 ⑧ 72 ⑨ 6 ⑩ 2
⑪ 9 ⑫ 8 ⑬ 40 ⑭ 36 ⑮ 14 ⑯ 54 ⑰ 25 ⑱ 27 ⑲ 56 ⑳ 32
㉑ 4 ㉒ 30 ㉓ 8 ㉔ 45 ㉕ 35 ㉖ 8 ㉗ 7 ㉘ 42 ㉙ 10 ㉚ 18
㉛ 16 ㉜ 9 ㉝ 63 ㉞ 36 ㉟ 15 ㊱ 12 ㊲ 12 ㊳ 8 ㊴ 9 ㊵ 72

ABOUT THE AUTHOR

Dr. Chris McMullen has over 20 years of experience teaching university physics in California, Oklahoma, Pennsylvania, and Louisiana. Dr. McMullen is also an author of math and science workbooks. Whether in the classroom or as a writer, Dr. McMullen loves sharing knowledge and the art of motivating and engaging students.

The author earned his Ph.D. in phenomenological high-energy physics (particle physics) from Oklahoma State University in 2002. Originally from California, Chris McMullen earned his Master's degree from California State University, Northridge, where his thesis was in the field of electron spin resonance.

As a physics teacher, Dr. McMullen observed that many students lack fluency in fundamental math skills. In an effort to help students of all ages and levels master basic math skills, he published a series of math workbooks on arithmetic, fractions, long division, algebra, trigonometry, and calculus entitled *Improve Your Math Fluency*. Dr. McMullen has also published a variety of science books, including introductions to basic astronomy and chemistry concepts in addition to physics workbooks.

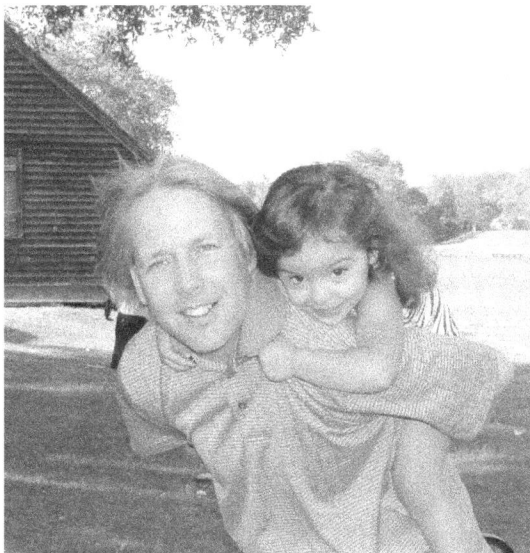

Author, Chris McMullen, Ph.D.

ALGEBRA

For students who need to improve their algebra skills:

- Isolating the unknown
- Quadratic equations
- Factoring
- Cross multiplying
- Systems of equations
- Straight line graphs
- Word problems

www.improveyourmathfluency.com

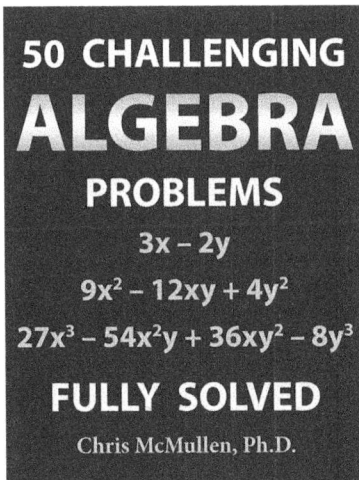

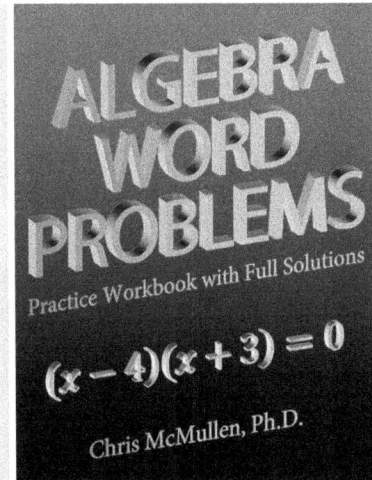

50 CHALLENGING ALGEBRA PROBLEMS
$3x - 2y$
$9x^2 - 12xy + 4y^2$
$27x^3 - 54x^2y + 36xy^2 - 8y^3$
FULLY SOLVED
Chris McMullen, Ph.D.

ALGEBRA ESSENTIALS PRACTICE WORKBOOK WITH ANSWERS
Linear & Quadratic Equations, Cross Multiplying, and Systems of Equations
$2x^2 - 3x = -1$
Improve Your Math Fluency Series
Chris McMullen, Ph.D.
Updated

ALGEBRA WORD PROBLEMS
Practice Workbook with Full Solutions
$(x - 4)(x + 3) = 0$
Chris McMullen, Ph.D.

MATH

This series of math workbooks is geared toward practicing essential math skills:

- Algebra and trigonometry
- Geometry
- Calculus
- Fractions, decimals, and percentages
- Long division
- Multiplication and division
- Addition and subtraction

www.improveyourmathfluency.com

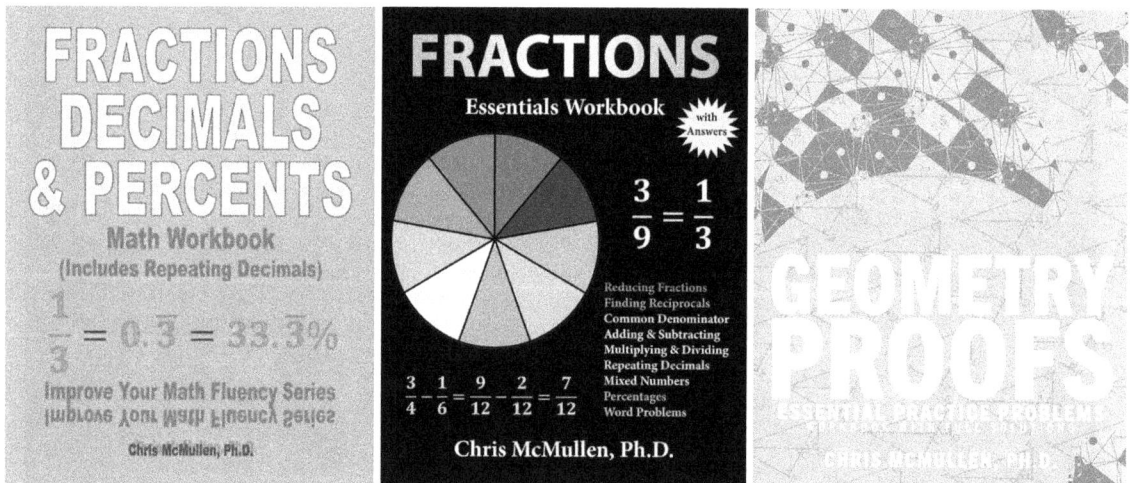

PUZZLES

The author of this book, Chris McMullen, enjoys solving puzzles. His favorite puzzle is Kakuro (kind of like a cross between crossword puzzles and Sudoku). He once taught a three-week summer course on puzzles. If you enjoy mathematical pattern puzzles, you might appreciate:

300+ Mathematical Pattern Puzzles

Number Pattern Recognition & Reasoning
- Pattern recognition
- Visual discrimination
- Analytical skills
- Logic and reasoning
- Analogies
- Mathematics

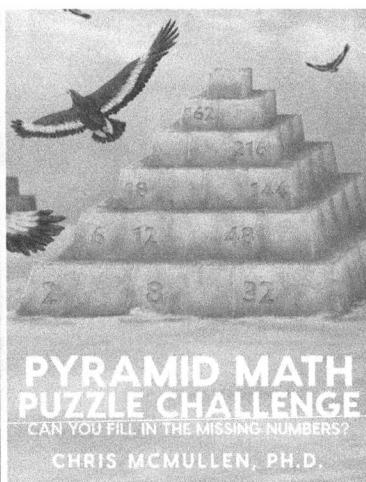

SCIENCE

Dr. McMullen has published a variety of **science** books, including:

- Basic astronomy concepts
- Basic chemistry concepts
- Balancing chemical reactions
- Calculus-based physics textbooks
- Calculus-based physics workbooks
- Calculus-based physics examples
- Trig-based physics workbooks
- Trig-based physics examples
- Creative physics problems
- Modern physics

www.monkeyphysicsblog.wordpress.com

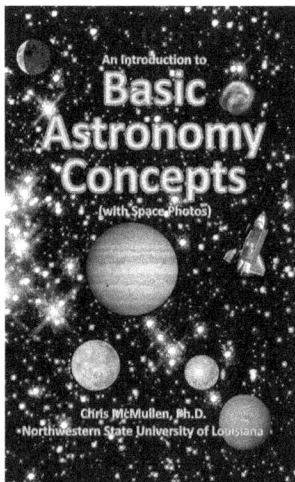

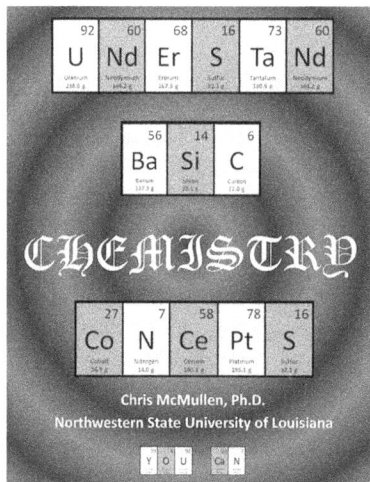

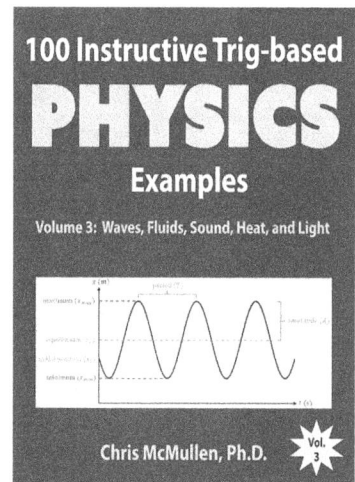

www.ingramcontent.com/pod-product-compliance
Lightning Source LLC
Chambersburg PA
CBHW081514040426
42447CB00013B/3225